Oklahoma Horizons Series

Oklahoma Horizons Series

Oklahoma Heritage Horizons Committee, 1985

W.R. Stubbs, *Chairman*
Henryetta

Lowe Runkle, *Vice-Chairman*
Oklahoma City

W. David Baird
Department of History
Oklahoma State University

William H. Bell
Tulsa

J.E. "Jack" Carter
Lawton

Reba Collins
Claremore

Jack T. Conn
Oklahoma City

Arrell M. Gibson
Department of History
University of Oklahoma

Jack Haley
Western History Collection
University of Oklahoma

Robert A. Hefner, Jr.
Oklahoma City

Henry C. Hitch, Jr.
Guymon

Earnest Hoberect
Watonga

Mrs. Patrick J. O'Hornett
Oklahoma City

Mrs. Jerry Richardson
Oklahoma City

F.A. Sewell, III
Clinton

Michael C. Thomas
Oklahoma City

A HISTORY OF THE INTERNATIONAL PETROLEUM EXPOSITION AND CONGRESS
1923-1979

Oklahoma Horizons Series

A HISTORY OF THE INTERNATIONAL PETROLEUM EXPOSITION AND CONGRESS 1923-1979

by
James P. Walker

Series Editor: Kenny A. Franks

UNIVERSITY OF TULSA - McFARLIN LIBRARY

Published for the Oklahoma Heritage Association by Western Heritage Books, Inc.

Copyright © 1984
Oklahoma Heritage Association

All rights reserved. No part of this book may be reproduced or utilized in any form or by any means, electronic or mechanical, including photocopying and recording, or by any information storage and retrieval system, without permission from the publisher

Manufactured in the United States of America for the Oklahoma Heritage Association, Oklahoma City.

ISBN: 0-86546-060-4

FOREWORD

The International Petroleum Exposition and Congress, affectionately called the IPE or the Tulsa Oil Show, the history of which is here presented, is a saga of an important segment of the Petroleum Industry, and also of civic involvement which established a city's position in a prospective growth industry. From an industrial standpoint, after the initial boost by civic non-oil men, it became the lengthened shadow of two entrepreneurial giants in oil and gas development, W.G. Skelly and W.K. Warren.

It also portrays the greening of an industry amid changing styles and modes of living. For example, exhibitors at early shows were willing to man displays for 10 to 14 hours a day and at night too, while latter-day shows were cut to four and three days, open only in daytime.

The IPE introduced the steel derrick, the truck-mounted well-servicing unit (which eliminated the forests of servicing derricks), and practically every landmark invention and development in equipment and services from the early 1920s for more than half a century.

To those with knowledge of the mentioned persons and events, Dr. Walker's history will be extremely interesting. To those of later generations, or late-comers to the oil and gas industry, it provides an insight into the competitive spirit and eagerness of industry suppliers to get their inventions, innovative designs, and improvements over to a growing industry faster than could be done by individual sales calls, word of mouth, etc.

Furthermore, the history is told, to a large extent, through inverviews with a few of the key characters who still survive—a luxury usually denied the evaluation of the annals of time except for autobiographies. Extensive research was undertaken into media and other IPE files of individuals and those which the University of Tulsa has in its Library Archives. Media files provide news of current events, although such is actually history "as it is being poured into molds." Thus we have, in a measure, both a condensed biography and autobiography of a segment of American entrepreneurship.

The IPE ran the gamut of big industries—from the take-off or entrepreneurial period through the incline development period, through the plateau period, which also inculded competition, and to the decline period. The decline period, in the IPE's case, involves failure to meet competition, failure to see the IPE as an industry event rather than a Tulsa event, and perhaps because the directors were involved in a crisis era after the OPEC price rise, and were simply becoming tired of it all.

Leslie Brooks
Tulsa, Oklahoma

PREFACE

"It's just obvious that the IPE played a major role in the history of Oklahoma. The whole world turned toward Oklahoma...for the IPE," said William H. Bell in an interview with the author in August of 1982. Bell, who was born and "partially raised" in the Far East, said that even there he heard of the IPE in his youth, and in his mind it became a kind of Mecca for those devoted to the adventure and hope of the treasures to be found at the end of a drill stem. For most of its 56-year run, the IPE was *the* only authoritative informational forum for the men who sought to unlock the mysteries of the earth's hydrocarbon riches. They poured into Tulsa, the Oil Capitol of the World, to see the latest marvels in equipment and supplies, to hear experts unravel new theories about finding and processing oil and gas, and to renew old friendships that began in an oil patch somewhere in the world. They came to salute King Oil and to find new and better ways of invading his realm.

This story of the International Petroleum Exposition and Congress is one that has to be told, and, as Bell noted, now is the opportune time while many of the first-generation entrepreneurs who conceived the great oil exposition are here to keep the story honest. Sadly, many are gone who could have given a glint of insight here or added an intriguing anecdote there, but many of their peers in a sense are the "co-authors" of this book and have represented their interests in the telling of the story.

What the International Petroleum Exposition and Congress was, it taught the worldwide petroleum industry to be: purveyors of knowledge, sharers of information, innovators in an industry whose lifeblood is innovation, colleagues in an enterprise which largely makes possible modern life as we know it, and all of this says nothing about the pure and simple good times provided by the IPE. There were "high old times" for American oil men, for international oil men, for Tulsa businessmen, civic leaders, and school kids. No doubt, millions of people remember the IPE as a "good time" who have little idea of its purpose. They would all know, however, that it had something to do with oil—and that knowledge was one of its purposes.

The IPE was also something else. Throughout its 56 years it was a marvel of non-profit enterprise which proved that an institution of worldwide significance could rise and stand without government subsidy and without returning a penny of profit to those who created and sustained it although it benefitted them all. It remains, perhaps because it eschewed subsidy and profit, the greatest showcase the petroleum industry has ever known.

What the IPE also proved each time its gates swung open was that change is a constant in the petroleum industry, and finally that process overtook the IPE itself. The rise of the new "glamor" segment of the industry—offshore exploration—had something to do with the demise of the IPE; the spread of regional expositions had something to do with it; and the decentralization of the oil industry which negated the claims of Tulsa to the title "Oil Capitol of the World" had something to do with it. But all of this simply says that the IPE worked itself out of business. What it exemplified and taught for 56 years finally "took," and different groups of entrepreneurs began applying its proven principles, not, however, with the triumphant success of the great Tulsa mentor.

My sincerest wish is that those who helped in the preparation of this book will be pleased with the form their contributions have taken in its pages.

Special thanks must go to Leslie Brooks, a guiding force who gave copiously of his time and attention to assure that the needed materials were available and that the right people were contacted. Significant contributions of time and guidance also were made by the following people: W.K. Warren, long-time IPE president and chairman of the Warren Medical Research Foundation; William Sherry, independent oil operator and leader of the IPE Old Timers' activities; William H. Bell, administrator of the Chapman-McFarlin-Barnard Interests; P.C. Lauinger, chairman of the board of Pennwell Publishing Company; John Houchin, retired president of Phillips Petroleum Company and the last president of the IPE; James O. Kemm, executive vice president of the Oklahoma-Kansas Oil and Gas Association and a leader in organizing the IPE film festivals; Otha Grimes, president of Portable Gasoline Plants, Inc., and of Grimes Gasoline Company; Herbert Fries, also of Portable Gasoline Plants, Inc., and a leader in the NOMADS international oil men's association; Randolph Yost, retired president of Amoco Petroleum Corporation and a former IPE president; Howard Hopkins former official photographer of the IPE; Carl Lawrence and his secretary, Ms. Joan Boyer, of Pennwell Publishing Company; Frank Stankey, Andrea Clark, and Kimberly Kirkpatrick of the Oklahoma Historical Society's research library; Bob Enright, editor of the *Oil and Gas Journal*; Drs. Paul Lambert and Kenny Franks of the Oklahoma Heritage Association for opening doors and making timely suggestions; and to Tony Walker for help in gathering photographs.

Thanks go also to those who helped in a variety of ways: Dr. Richard Madaus, director of the Northeastern State University library; Ms. Victoria Sheffler and Ms. Delores Sumner, Special Collections librarians at Northeastern State University, and to their counterparts in the Special Collections Division of the University of Tulsa; Charles Rogers; my colleagues Drs. Kirk Boatright, Tom Cottrill, and Lanny J. Reed; my friend Dr. Odie B. Faulk, who pointed me toward the project—or rather pointed the project toward me—and who served as my mentor throughout the work; Dr. Charles Prigmore, vice president of Northeastern State University for crucial help near the deadline; and to President W. Roger Webb of Northeastern State University, who supported the work with both funds and encouragement.

Finally, I thank those who suffered for the cause: my wife Lori for many late nights spent proofreading my often garbled manuscript, and my children, Matthew, Andrew, and John, who tolerated their parents' preoccupation reasonably well.

Readers of this book should find what they are looking for—an accurate and balanced history of the International Petroleum Exposition and Congress. If they do, they should thank the many people who generously gave their valuable time to assist the author in gathering materials; if they do not, the culprit's name is below.

James P. Walker
Tahlequah, Oklahoma

CONTENTS

FOREWORD	vii
ACKNOWLEDGEMENTS	ix
PART I: THE BEGINNINGS	3
1923	3
1924	10
1925	15
1927	20
1928	24
1929	28
1930	31
PART II: AFTER THE DEPRESSION AND BEFORE THE WAR	36
1934	36
1936	46
1938	50
1940	56
PART III: BACK TO BUSINESS	65
1948	65
1953	74
1959	81
PART IV: FINAL YEARS	90
1966	90
1971	96
1976	101
1979	106
EPILOGUE	110
NOTES	113
BIBLIOGRAPHY	120
INDEX	121

ILLUSTRATIONS

Tulsa in 1905	4
Judge S.H. King	5
First IPE Board of Directors	7
Hotel Tulsa	9
IPE gates in 1923	10
Invitation to the second IPE	11
Princes Oklahoma and King Petroleum's float	12
Drawing for prizes, 1924	14
W.G. Skelly	16
Skelly Oil Company invitation to the IPE, 1925	18
Advertisements in 1925	19
Outside exhibits, 1927	21
W.G. Skelly's opening address, 1927	22
Scenes in 1927	24
Exhibit booths and buildings, 1928	25
Main street of the IPE, 1928	26
Springpole drilling exhibit	28
E.B. Beeser speech, 1929	29
The gate at the 1929 IPE	31
Visitors inspecting exhibits, 1929	32
Quigley Company display	34
William B. Way	35
Aerial view of the IPE, 1934	37
Welcoming ceremonies, 1934	39
Scenes in 1934	40
Marathon Oil booth, 1934	41
Pump engines, oil memorabilia, 1934	43
Safety demonstration	44
E.W. Marland	47
Opening ceremonies, 1938	52
W.G. Skelly and participants, 1938	55
German representatives, 1938	56
IPE grounds, 1939; aerial view in 1940	57
Keynote speech, 1940	58
W.G. Skelly in 1940	59
Andrew F. Schoeppel	60
Visitor scenes, 1940	61
California Oil World booth, 1940	62
Aerial view, 1940	64
Aerial view, 1948, and W.G. Skelly	67
IPE officials and guest speakers, 1948	68
Model of rotary rig	69
Map showing flying time to Tulsa	70
Technical exhibit, 1948	71
Centrifugal compressor; W.G. Skelly and Perry L. Nunley	73
Oil valve	74
The 1948 IPE	75
Opening ceremonies, 1953	76
Charlton H. Lyons	77
U.S. Steel display, 1953	78

Lois DeFigh	79
Johnston Murray and W.G. Skelly	81
W.K. Warren	82
IPE office, 1959	84
Oklahoma Building	85
National Supply Company exhibit, 1959	86
IPE grounds in 1959	87
Joe McQuire and security in 1959	88
Foreign visitors in 1959	88
Extending invitations to foreign visitors	89
Martin Dwyer	90
Tulsa Exposition Center	92
Henry Bellmon congratulates IPE officers	92
The Golden Driller	93
Exhibits in 1966	94
Products available, 1966	95
Planning for the 1971 IPE	98
Leslie Brooks	99
Trailer-mounted rig, 1971	99
Monument commemorating the IPE	102
Robert Sharbaugh and William J. Sherry	103
Engineers' Day Luncheon, 1976	104
Invitation to foreign delegates	105
Desk & Derrick Club presidents	105
W.K. Warren, Randolph Yost, John Houchin	106
Invitation, 1979	107
Mack Truck exhibit, 1979	108
Planning for 1979	109
Petition for dissolution of the IPE	110
John Houchin presenting checks to universities	111

A HISTORY OF THE INTERNATIONAL PETROLEUM EXPOSITION AND CONGRESS
1923-1979

CHAPTER ONE
THE BEGINNINGS
(1923-1930)

When the International Petroleum Exposition and Congress, or IPE, began in 1923, it truly was a show—with a king and queen, showgirls, vaudeville acts, sideshow barkers, and all kinds of ballyhoo. In its early years the IPE only remotely resembled—in attitude, atmosphere, and setting—the sophisticated industrial showcase it became. But the planners of those early shows had no standard by which to measure their enterprise, for nothing like it existed in the oil industry. There was much last-minute scurrying about to assemble machinery and arrange displays in the tents and sheds huddled around Tulsa's old Convention Hall, but there always was time for a cup of coffee or for talk with old friends, even if it meant that an exhibit had to be finished as crowds of spectators circulated in, around, and sometimes through the makeshift display booths.

What everyone looked forward to was the arrival of King and Queen Petroleo's parade from downtown with their entourage of bands, dignitaries, and boys on bicycles. The show was not really "on" until these monarchs arrived. Serious business and "high old times" were very nearly of equal importance during these early years. IPE officials, exhibitors, and visitors were all aware of this, and they heartily approved.

After the first exposition, the consensus was that everyone had had such a good time that the show should go on but that it should be bigger and better. Tulsa's congested downtown area could not accommodate expansion of the show, so it was moved to the old Barton Show Grounds and, ultimately, to a 16-acre plot rented from the Tulsa State Fair. There the IPE began to take on the appearance of the permanent institution it became. Buildings began to rise and drill bits began to bore demonstration wells into the east Tulsa soil. The International Petroleum Exposition and Congress took root.

1923

Whistles, locomotive bells, and trumpets heralded the arrival of King Petroleo at Tulsa's Frisco Railway Station at 9:45 a.m. on October 8, 1923. A special train had been provided for his trip from the little town of Dawson just east of Tulsa, and he was welcomed by a throng of Tulsans and visitors to the first International Petroleum Exposition and Congress. Preceded by a band playing, "Yes, We Have No Bananas," 79-year-old S.H. King, in the role of King Petroleo, made a regal journey to the exposition site. A native Virginian who had been chosen monarch at least partly because he was one of the first attorneys to represent the Osage Indians, King had helped usher in a remarkable era in the history of Oklahoma business and industry, the 56-year run of the IPE. Nor can it be ignored that this era, at least partly because of the IPE, marked what so far has been the most remarkable period in the growth and development of the city of Tulsa. As Leslie Brooks, former advertising executive whose firm handled IPE publicity, has remarked, "I would say that the show was an important factor in helping Tulsa become the Oil Capitol of the World. It certainly was a booster to it, although it didn't start it off."[1]

Curiously, many Oklahoma oil men, at the opening of the IPE in 1923, believed that their state had almost exhausted its promise as an oil producer inasmuch as the great fields of Glennpool, Cushing, and Healdton had faded into history.[2] However, there still were those who tenaciously believed that for Oklahoma the best was yet to come and that the oil industry in the Sooner State should join the large number of industries across the country that were showcasing their equipment, supplies, products, and processes through industrial shows and fairs.

One exhibit featuring the oil industry already had been held in Tulsa, although it was miniature in scale compared to the IPE. The Tulsa Purchasing Agents Association, under the leadership of its secretary, O.V. Borden, had presented an exhibition at its headquarters on south Cincinnati Avenue and had offered rental space where oil

3

Tulsa, the future Oil Capitol of the World, looking south on Main Street in 1905. Courtesy Beryl D. Ford.

field supply firms could exhibit their latest wares. When it opened in 1922, this show, with its limited indoor exhibition of oil field supplies and equipment, was very successful and had been attended by many visitors.³ The most important person among them, so far as the IPE was concerned, was Tulsa attorney Earl Sneed.

Impressed by what he saw, Sneed conceived the idea of organizing an exhibit for the entire oil industry and on a much larger scale. He elaborated on this concept in an editorial in the Tulsa *American Saturday Night* of March 3, 1923: "An International Petroleum Exposition and Congress with all its sideshow features would give thrills to the young people, knowledge to the oil fraternity, opportunity to make worldwide acquaintances, renew friendships, and firmly establish Tulsa for all time to come as the oil center of the entire world."⁴

At the end of his editorial Sneed invited "all manufacturers of oil field equipment and refinery supplies, and all those interested in production…to form such an organization."⁵ He went on to point out that an event of this kind would "be the first time in the history of petroleum that an exposition, international in character and designed and planned for the oil industry and those directly dependent upon it has ever been held."⁶

He further added that the kind of exposition he was proposing would include "all branches of the oil industry" and would allow oil men to "get together with the spirit of rivalry absent to discuss problems which affect the good of the entire fabric of oil, and allied industries."⁷

Sneed made this suggestion to the Tulsa Chamber of Commerce in a letter to A.L. Farmer, a member of the Chamber's board of directors. Farmer presented the letter to the Chamber, and, after some discussion, the directors voted to appoint a committee to study the feasibility of establishing an industry-wide oil exposition in Tulsa.⁸ The members of this committee were J. Burr Gibbons, H.O. McClure, and William Holden.⁹ These men contacted everyone they knew in the oil industry in the Tulsa area and also met with the city's leading merchants and property owners in an effort to learn how much support could be generated for the proposed oil show. Leslie Brooks later would recall:

> Interestingly, the show as actually promoted more by merchants than by oil men. There were oil men involved, all right, but the merchants actually put some money into the venture, and some of this was not paid back until the 1930s. I think that one of the most im-

4

Judge S.H. King, the first King Petroleo. A native Virginian, the 79 year-old King was chosen partly because he was one of the first attorneys to represent the Osage Indians. Courtesy Leslie Brooks.

portant points about this oil show is there never was a bit of government or city money in it. It was a private venture all the way.[10]

The committee of Gibbons, McClure, and Holden reported back to the Chamber that its investigation had revealed sufficient interest in an oil show that a public meeting should be held to lay the matter out before the entire Tulsa business community. This meeting was held on May 11, 1923, at Convention Hall at 105 West Brady Street. It was attended by a "good cross-section of Tulsa's business community, oil industry, and civic associations."[11] At this meeting it was proposed and unanimously approved that a corporation be formed under the laws of the State of Oklahoma to be known as the International Petroleum Exposition, Incorporated, and that this be governed by an 11-member board of directors.[12]

This first governing board was made up of some of the "leading lights" of Tulsa's business and civic community, including: Lewis B. Jackson, an independent oil operator; J.M. Hayner, an oil executive; William Holden, secretary of the Tulsa Chamber of Commerce; James J. McGraw, president of Exchange National Bank; A.V. Bourque, a newspaper reporter; Alf G. Heggem, an oil equipment manufacturer; W.A. Melton, general superintendent of Liquified Petroleum Gas Company; James H. Gardner, president of Gardner Petroleum Company; Thomas J. Hartman, president of Producers National Bank; and H.O. McClure, president of the Tulsa Chamber of Commerce and Atlas Life Insurance Company.[13]

In addition to the directors, a general committee of more than 75 Tulsans was organized into 12 working committees: finance, attractions, exposition, convention, parades and pageants, transportation, scientific and technical exhibits, public safety, auditing, entertainment, decorations, and building and grounds.[14]

The membership of these committees was a "who's who" of Tulsa's leading citizens: L.D. Armstrong, A.L. Beekly, Carl Blackman, A.V. Bourque, O.V. Borden, A.F. Bourne, F.W. Bryant, D.E. Buchanan, John Champion, Fred S. Clinton, M.D., O.L. Cordell, E.H. Cornelius, A.B.C. Dague, W.E. Espy, C.T. Everett, Charles F. Fallen, E.R. Filley, T.M. Pariss, N.R. Graham, R.D. Gwynne, J. Burr Gibbons, Frank Glasscock, E.B. Guthrie, R.L. Ginter, H.H. Goddard, J.H. Gardner, J.M. Hayner, William Holden, Alf G. Heggem, T.J. Hartman, A.W. Hurley, Frank Hinderliter, Summers Hardy, Richard Hughes, A.C. Holmes, W.R. Hamilton, R.P. Humes, L.B. Jackson, Cornelius Kroll, L.E. Kennedy, I.G. Long, W.L. Lewis, C.M. Lemason, J.J. McGraw, H.O. McClure, W.A. Melton, P.M. Miskell, Charles Meyers, Everett Manning, J.H. McBirney, D.W. Moffitt, G.L. Matson, R.L. McFarland, T.F. Mayer, C.M. Murray, H.E. McElroy, J.S. McKelvey, Hollis F. Porter, Asa E. Ramsey, L.G. Rosser, E.A. Richards, Ralph C. Riley, Harold E. Roe, W.R. Ritchie, Harry Smith, J.A. Sartori, E.T. Tucker, J.A. Udden, W.A. Vandever, M.M. Valerius, E.H. Wiet, Allan Whiteside, J.S. Warren, W.M. Welch, J.E. Minger, and John Zink.[15]

These leaders set about planning the first

"greatest oil show on earth." Shortly after incorporating the IPE, these planners added "and Congress" to the formal title of the exposition, although not to the formal title of the organization. The addition of "and Congress" to the title solved several problems. Federal law required expositions and conventions to include some educational feature before allowing reduced railroad rates to those attending; by adding guest speakers and presenting technical papers the planners felt they would attract more out-of-town visitors.[16] The founders reasoned that the gathering would provide "a meeting place where problems of the oil industry could be discussed in true cooperation by members of all branches of the industry and solutions found."[17] Moreover, they hoped this would attract foreign interest, and from the beginning they invited representatives and speakers from abroad.

As preparations for the first IPE continued, the directors named Lewis B. Jackson from their number as president. The other officers included: J.M. Hayner and W.A. Vandever, vice presidents; J.J. McGraw, treasurer; and W.A. Holden, secretary. The officers, in turn, hired Edward P. McIntyre, a veteran director of shows and carnivals, to manage the first show. Tulsa's Convention Hall and the area surrounding it was chosen as the site, and the directors asked city authorities for permission to cordon off the surrounding five blocks.[18]

The next step was to provide facilities at which exhibits would be housed without being exposed to the weather, and Braden Construction Company, which had submitted the low bid, was given a contract to erect the booths and covered buildings needed. Also to be constructed in a vacant lot directly behind Convention Hall was a structure to house the scientific, technical, and historic exhibits. Preparations moved along so smoothly that on September 7 General Manager McIntyre was able to inform the directors that all available exhibit space had been reserved and that applications for additional space were still coming in.[19]

Again the directors approached city officials to ask that additional space be cordoned off, to which city officials eagerly agreed, allowing another block for the exposition. The Braden Company was put to work again to construct the necessary exhibit facilities. Convention Hall itself was reserved for the variety show and the other meetings and conferences planned by various petroleum associations and organizations.[20]

The goal of the IPE board was to make the exposition area completely self-sufficient. This required a massive effort to provide adequate facilities for food, rest, and communications. When the preparations were done, the exposition area had become a city within a city.[21]

Several late decisions were made concerning entertainment and pageantry in connection with the exposition. in order to encourage greater involvement by Tulsa's citizens, the directors decided to stage two parades, one for King Petroleo and the other for merchants. A related activity was a beauty pageant for choosing Queen Petroleo, who would ride with the king in his triumphal parade from the Frisco Railroad Station to Convention Hall; there the king and queen would reign from October 8 through 14, 1923, during the International Petroleum Exposition and Congress.[22]

Just before this spectacle began, the directors made a decision which proved crucial in light of later developments. They decided that, given the expenses incurred in preparing the exposition, they should hedge their bet by taking out an insurance policy against a rain-out of the parade and other activities. President Jackson on October 1, 1923, appointed Dr. Fred S. Clinton chairman of a committee to arrange for rain insurance. This committee routinely discharged its responsibility, and work continued on preparations for the oil show. Later this routine action proved more important than any other the directors took.

Once the policy was purchased, all thoughts of misfortune were put aside, and everyone plunged into last-minute preparations for housing, entertaining, and feeding the more than 100,000 visitors expected to attend the first IPE. Intensive efforts were made to "get out the word" about the show. An aggressive advertising campaign was undertaken by the directors, and Tulsa newspapers fully cooperated with the effort.[23] The *Tulsa Daily World* and *Tulsa Tribune* explained that the purpose and goals of the exposition were:

> ...to promote a feeling of cooperation between large and small producers, refiners, manufacturers and marketers...; to provide a place where the manufactured articles, tools, supplies, machinery, and accessories used in the industry may be exhibited to the crowds at-

The first IPE Board of Directors. Top row, L to R: T.J. Hartman, J.M. Hayner, W.A. Vandever, James H. Gardner. Middle row: H.O. McClure, L.B. Jackson, W.A. Melton. Bottom row: Alf G. Heggem, William Holden, Edward McIntyre, James J. McGraw, A.V. Bourque. Courtesy Oil and Gas Journal.

tending the Exposition and Congress, their good points demonstrated and the services offered by each exhibitor explained to the prospective customers. The members of the organization hoped that the meeting would also hasten the standardization of tools and equipment and thus eliminate one of the great sources of waste and inefficiency...in the industry...; to promote in the oil field workers a sense of pride in their work by encouraging rapid and efficient performances of their tasks; to attract oil men in all walks of life from all parts of the world; [to provide] an annual reunion of old friends and a meeting place for new acquaintances, [and to] educate the public [so as] to eliminate fake promoters from the oil business and present the true facts concerning the industry....[The show also would promote] the formation of the nucleus of a Museum of Petrology...[and bring together] articles of historic value showing the development of the industry from its beginnings, as well as recognize and reward distinguished service on the part of oil field employees by the awarding of a medal or medals annually for the performance of outstanding actions in saving life or property or such other acts.[24]

Aerial view of the buildings and grounds of the first IPE (Boulder and Brady streets). Tulsa's old Convention Hall can be seen in the upper left (marked "Auditorium"). Courtesy Leslie Brooks.

With these lofty ideals, the IPE was ready to begin. Headlines in Tulsa's two daily newspapers reflected the mood of the city as the show prepared to open:

> International Petroleum Exposition, For Progress and Improvement of the Industry Through Education, Cooperation, Standardization; Grand Reunion, Homecoming of Oil Fraternity; Acres of Exposition Buildings and Exhibits; Gorgeous, Spectacular Pageant Petrolia; World's Hippodrome, Ten-Act All-Star Show; Reduced Fares on All Roads; Tulsa, the City Will Astonish You; Inspiring Skyline; Tremendous Concentration of Refining Industry; City of Over 100,000 Built Largely in the Last Decade; Many Conventions in Tulsa the Same Week; The Many Weeks of Preparation Are Over; On With the Show.[25]

The show began on Monday, October 8, with King Petroleo's Parade. The weather cooperated with bright, clear skies as the parade coursed the 26 blocks from Second and Detroit streets to Archer and Cheyenne streets. The parade was splendidly planned and bountifully appreciated by Tulsans as routine activities in the city came to a halt to allow everyone to participate in the greatest oil extravaganza known before or since. Many of those watching were students, for both the public schools and the University of Tulsa had dismissed classes.[26] At a gala ball at the Tulsa Club the previous Saturday, Miss Dorothy Vensel had been selected Queen Petroleo. She was attended by dutchesses Ann Kennedy, Lillian Randall, Katherine Gavin, Nell Cook, Rosaline Hollow, and Katherine Reisling.[27]

The festivities featured twice-daily performances of the World's Hippodrome at Convention Hall. These performances were much like those of other vaudeville extravaganzas of the period:

> The senational dance team of Vallal and Zermaine from the Paris Follies headlined the extravaganza. Miss Vallal was renowned as one of the dancing partners of Rudolph Valentino in the motion picture, "The Four Horsemen of the Apocalypse." Art Landry's band was advertised as having real solo artists, jazz, and classical. Finks' Comedy Circus was a comedy animal act featuring a bucking mule named Spot. The six Belfords were tumblers and balancers, and the Ballad of Jewels led by Miss Hazel Wallak supported Vallal and Zermaine.[28]

A more serious entertainment was offered by the *Oil and Gas Journal*. It sponsored the showing at the Majestic Theater of a film specially produced by the Sinclair Consolidated Oil Company, "The World Struggle for Oil," which graphically illustrated how oil had been formed down through the eons. This film was one of the most popular events of the exposition with the turnout for it heavy at each showing. Admission was gained by presenting a ticket that could be obtained at the *Journal*'s exhibition booth on the exposition grounds.[29] The *Journal* contributed in other ways to the success of the IPE, as P.C. Lauinger later would recall:

> We were a promotional factor for them. That is, we recorded the event for the purpose of promoting the exposition. Our readers want to be told the whole story, like other readers.... What we did was just tell the story each year..., and that was our part in the whole business. We always had a special edition devoted to the exposition, and for a good while it was the largest edition of the year. As the exposition grew bigger, so did our exposition issue.[30]

The first International Petroleum Congress took place on October 9 and 10 in Convention Hall. E.R. Perry, vice president of Cosden and Company, took charge of the session on October

9, and W.H. Grey, president of the National Association of Independent Oil Producers, conducted the meeting on October 10. Outside the hall the scene was far less sedate:

> In one area of the grounds were eleven kinds of oil field transportation devices (tractors, trucks, tool racks, tank trucks, and automobiles). The booth of the *Oil and Gas Journal* in the center of the exposition was equipped with telephone and telegraph facilities. Boy Scouts served as volunteer messengers and guides. Cosden and Company exhibited a paraffin wax model of the Washington Monument....A War Department searchlight installed and operated by Public Service Company and reputed to be the largest ever used in the southwest beamed forth from atop a standard steel rig.... Sophisticated young flappers with bobbed hair marveled at the Chinese-delicate diamond drills, gas masks, Indians and oil derricks. The science and technical exhibit occupied an entire building immediately behind Convention Hall. This display was probably the most complete and best organized at the first of the Exposition. The exhibiting firms' materials were divided into the six major divisions of the oil industry: exploration, refining, transportation, production, marketing, and consumption.[31]

Included among the technical displays were the tools used by Colonel E.L. Drake in drilling the country's first commercial oil well near Titusville, Pennsylvania, in 1859. Many other priceless artifacts from the beginnings of the petroleum industry in America also were displayed.

Oil was king and those who served it were princes and princesses during the second week of October in 1923, and Tulsa became Camelot. The city had a population of just over 110,000 that year, a number swelled by several thousand as visitors and oil personnel streamed into the city. The first housing committee of the IPE, headed by J.E. Sharpe, scrambled to find sufficient lodgings for the throngs of visitors and dignitaries, resorting finally to rooms in the homes of private citizens who offered to help. Through the efficient efforts of this committee, everyone was provided with adequate lodgings.[32]

Everything went as planned during the first four days of the IPE. The weather was satisfactory, and the exhibits and entertainments were

The Hotel Tulsa, headquarters and meeting gplace for IPE guests and dignitaries during the early years of the show. Courtesy Beryl D. Ford.

spectacular. A few complaints were heard that the city had not responded with sufficient enthusiasm and that the oil industry itself had contributed inadequately. For the most part, however, this fledgling exposition was accorded a success and given a hopeful prognosis, although during the final three days of the show rain clouds rolled in and an unrelenting downpour began. Many exhibits were ruined, and a number of events had to be canceled.[33]

Nevertheless, the climactic event of the show was carried through despite incessant rain. As an educational exercise, Frank Hinderliter and Oscar C. Staples had undertaken to drill an actual oil well on the IPE grounds during the exposition, and it was announced that the well would be "shot" on Saturday, October 13, at 1:00 p.m. and that a large crowd was expected.[34] Other wells had been drilled during the exposition, but the "shooting" of this one had dramatic overtones that stirred much interest. When Saturday afternoon came, a large crowd gathered only to be told that the "shooter," who had been sent by the Eastern Torpedo Company, had bogged down on muddy roads and had been unable to reach Tulsa. The event was postponed until the next evening.

On Sunday afternoon W.H. Swartz, assistant manager of the Eastern Torpedo Company, strode to the well bore, measured its depth, and prepared a charge of nitroglycerin. After assuring himself that the spectators were safely out of range, he ignited the explosive with an electrical charge. Tulsa's cloud-darkened sky lit up for an instant with a brilliant flash followed immediately by the sound of fluid spewing out of the well bore under high pressure. Many bystanders were doused by the mixture of mud and water that shot up from the well, but this did not dampen the high spirit of excitement that surrounded the event.[35] This explosion brought the first IPE to a close—but signaled the birth of an enterprise that would become a chapter in the success story of American commerce.

The *Tulsa Tribune* of October 1, 1924, from the perspective of an intervening year, summarized the success of that first oil show and, although frowning slightly at certain excesses of "hoopla and ballyhoo," brought in a favorable verdict. The IPE would continue successfully in the city, said the *Tribune* article, because Earl Sneed's major suggestions had been carried out and the first show had much that had appealed to the general public. The IPE had earned the confidence of Oklahoma's citizens, especially Tulsans, said the *Tribune*, adding that it had attracted foreign interest, marking "the first time in the history of the petroleum industry that an exposition, international in scope, designed and planned for the oil business and those indirectly related to it had been held on an industry-wide basis." The article concluded that the oil industry had acknowledged that Tulsa was the logical place for the show because of the large number of companies headquartered there and because it was the geographic hub of the Mid-Continent territory.

1924

Edward McIntyre, IPE general manager, submitted his report on the 1923 show to the board of directors in November, nearly a month after Frank Hinderliter and Oscar Staples had closed the first exposition by shooting the well dubbed the Chicken Farm No. 1. The main theme of McIntyre's report was that expenses had to be reduced and that a site more suitable than "public streets" had to be found for the show. The answer, as McIntyre saw it, was to find a permanent location for the show in order that the grounds and exhibit facilities could be prepared "months in advance" of the exposition and so that long-term

Tulsa's populatin of 110,000 in 1923 was swelled by thousands as visitors swarmed through these gates to particpate in the first IPE. Courtesy Oil and Gas Journal.

agreements could be made with exhibitors "who would erect their own buildings."

McIntyre added that he had done some checking on possible locations and had concluded that the best possible site would be the old Barton Show Grounds at Peoria Street and Admiral Place. He urged this location because its 10 acres would provide ample space for large exhibits and because it was easily accessible from anywhere in town. He had even calculated the walking time from the Hotel Tulsa, where most out-of-town visitors and exhibitors stayed, to be 18 minutes.[37]

The directors did not immediately accept McIntyre's recommendation although they also were aware of the need for a new location. Several directors previously had favored moving the show to Central Park, across from Longfellow Elementary School on Sixth Street. However, this site ultimately was rejected apparently because of problems of parking and transportation. Another alternative was the old plant site of the Prairie Oil and Gas Company across Peoria Street to the west from the Barton Show Grounds. The show grounds, it was reasoned, would provide an ideal parking facility for IPE visitors and participants. The Prairie Oil and Gas propery finally was selected, and arrangements were made to reserve the show grounds for parking.[38]

As planning for the 1924 show began, one of the first decisions was to purchase another rain insurance policy. The rains of the previous year had shown the wisdom of such a move, for, as Leslie Brooks later noted, "The show would probably never have been held again if they hadn't taken out rain insurance. This insurance bailed the first show out financially."[39] The policy purchased in 1924 would pay $10,000 for each day the exposition was rained out; if the entire show was rained out, the IPE Corporation would collect $100,000 to meet its expenses.[40]

The publicity for the second IPE, as for the first, was keyed to an Indian theme inasmuch as Indian and oil history were inextricably linked in Oklahoma. The first oil well in Indian Territory had been drilled in 1859 in a search for salt by Louis Ross, brother of Cherokee Chief John Ross. The famous Red Fork Pool had been brought in on land leased from the Creek Nation and had led to a burst of development in what had been an Indian trading village southwest of Tulsa. Thus it was fitting that the IPE should celebrate the re-

The Oil and Gas Journal's invitation to the second IPE. Publicity for this exposition was keyed to an Indian theme. Courtesy Oil and Gas Journal.

lationship between oil and Oklahoma's Indian nations. A special Indian Night was designated for the show of 1924, and an Indian show was a featured attraction of the exposition entertainments.[41]

The 1924 IPE opened on Thursday, October 2, with a gala parade billed as the "Pageant of Princesses," and on October 7 a merchants' parade was held. From the beginning Tulsa's merchants were ardent supporters of the IPE. According to Leslie Brooks, "The show was actually promoted more by merchants than by oil men in the early years of the exposition."[42]

During the first IPE, professional entertainment had proven so successful that the directors planned a similar offering for the 1924 show. Circuit performers were brought in and billed as "Styles and Smiles of 1924," a vaudeville revue. This proved highly popular and financially successful.[43] A chief feature was the tent theater built to house this. The theater was decidedly more than a mere circus tent—it was a true theater, includidng 16 large dressing rooms, complex lighting and scenery, and an orchestra pit above which a large runway extended into the auditorium where

Princess Oklahoma, one of 12 "oil princesses," who took part in the Pageant of Princesses Parade before the opening of the second IPE. Courtesy Oil and Gas Journal.

King Petroleum's float in the Pageant of Princesses Parade in 1924. Courtesy Oil and Gas Journal.

there was room to seat 1000 to 1500 spectators. The canvas used in the tent was specially treated to resist moisture in order to protect both the audience and the elaborate costumes and scenery. This theater itself became one of the show's spectacles.[44]

Before the Wednesday evening performance of "Styles and Smiles," a new Queen Petrolia was named to inherit the crown from Dorothy Vensel. Romona Marcella Trees of Winfield, Kansas, daughter of an oil man, was selected from among 12 candidates who represented the oil-producing states.[45] All 12 of the young ladies appeared in the Pageant of Princesses Parade, which wound through 20 blocks of downtown Tulsa to the cheers and applause of 40,000 spectators. Thousands then trouped to the exposition grounds following the parade to enjoy the exhibits and attend the 2:15 performance of "Styles and Smiles." Such throngs attended the exposition that the directors feared an auto-pedestrian accident on Peoria Avenue; to prevent this, they had an elevated walkway constructed to carry foot traffic from the parking area at the Barton Show Grounds to the exhibit area.[16]

During the IPE show of 1924, excitement and high spirits permeated the atmosphere, but the exposition's serious purpose was not slighted. Renewal of professional and personal associations, one of those purposes, was exemplified by the activities of the Fraternal Order of the Knights of the Derrick, a charitable and social organization which had been formed just before the opening of the 1923 show. This organization sought to provide a forum where oil men could enjoy fellowship and discuss matters of importance to the worldwide oil fraternity. This group met in the exposition theater on the morning of October 4 and seriously discussed a proposal to found and endow a home for oil men who, because of age or infirmity, no longer could function productively in the industry. This proposal never bore fruit, but it indicated the strong fraternal feelings that existed—and still exist—among those who gamble their energy, skill, talent, money, health, and, sometimes, their lives in the always-risky quest for oil.[17]

The 1924 Congress was held in the exposition theater during three days of the show—October 6, 7, and 8—meeting each day at 10:00 a.m. Both American and foreign leaders in business, industry, and government were among the featured speakers. The president of the United States Chamber of Commerce sounded a major theme of that time in his address, making, as his title implied, "The Case for Business." M.E. Trapp, the governor of Oklahoma, reinforced this theme in his speech, "Business in Government," commenting that the government should stay out of business and that, to promote this end, the president's cabinet should be reduced, a sentiment that many American businessmen continue to share. Other speakers included George Smith, director of the U.S. Geological Survey, who spoke on "Needed—A Program for Profits"; Major General Mason M. Patrick, Chief of the Army Air Service, who lectured about the need for a uniform grade of aviation gasoline; A. Serebrovsky, president of the

Russian Oil Trust, who gave an overview of the Soviet oil industry; and Alfred Wirth, a German manufacturer of oil well equipment, who lectured on a wide spectrum of topics.[47]

A spectacular feature of the second IPE was the appearance of the TC-5, the Army's giant dirigible, which flew over the exposition grounds at 2:00 p.m. on October 10 to salute the city and the oil show.[49] This airship made several flights that day, on one of which Queen Petrolia accompanied the crew as they floated over the teeming exposition grounds and busy city. At the end of the day the dirigible lumbered toward its mooring at Wichita, Kansas. This marked the first visit to Tulsa by a dirigible and moved the *Tulsa Tribune* to remark, "If the hands of Father Time's clock in Oklahoma could have been moved back three or four decades today, the hills around Tulsa would have been dotted with Indian signal fires to herald the approach of a strange bird into the realm of the redmen."[50] The "strange bird" was only slightly less remarkable to those who watched it in 1924 than it would have been to "redmen" decades earlier.

Not only did the oil show attract strange flying machines, but also it brought a larger number of participants and visitors from other states and lands. Word had spread that the Tulsa exposition was *the* place to gather the latest information about procedures and techniques and to see the newest in supplies and equipment. Visitors from California, Pennsylvania, Arkansas, and Illinois joined with delegations from England, Russia, India, Germany, China, Peru, Colombia, France, Italy, Poland, Rumania, Japan, Mexico, Egypt, Venezuela, and Guatamala. The German and Russian delegates, Alfred Wirth and A. Serebrovsky, not only came as visitors but also as participants in the IPE Congress.[51]

There were three times as many exhibitors requesting space in 1924 as in 1923,[52] and a wider variety of oil field interests was represented. The *Tulsa Tribune* noted that the number of instrument companies exhibiting in 1924 had quadrupled over the number exhibiting the previous year.[53] In just two years the IPE had become a one-of-a-kind premier showcase for the petroleum industry.

And the city of Tulsa went "all out" to make IPE visitors and participants feel welcome and comfortable. Leading business and professional men organized into 11 teams of 13 to 15 members each to welcome visitors to the IPE. The Kiwanis and the Chamber of Commerce also threw their resources behind the effort at hospitality and provided teams of members to meet visitors.[54] Tulsa's business community was determined to create an atmosphere of hospitality and cooperation. Already the "Oil Capitol" was known as one of the friendliest cities in the Southwest, a place to which visitors could look forward to returning.

When visitors arrived at the exposition grounds in 1924, they found 2.5 miles of exhibits and demonstrations.[55] Those attending the IPE for the second time could see the significant growth that had taken place in just one year. Indeed, the exposition, "featuring more than ten million dollars' worth of displays supplied by 461 exhibitors,"[56] had expanded to almost twice the size of the first show. Two of the most popular exhibits were recently developed processes for producing gasoline; one was a method for extracting it from natural gas, and the other was a bubble tower and barrel that allowed the breakdown of heavy oils to derive gasoline from crude.[57] Also shown was a chemical developed by the Foamite-Childs Company which extinguished oil tank fires.[58] But the most spectacular exhibit was that of the National Supply Company; this incorporated four completely operational—and operating—oil rigs and covered more than 5000 square feet.[59]

At least one major exhibit was not directly oil-related, although it was of crucial importance to the continued development of the Oil Capitol. This was the Chamber of Commerce's model of Tulsa's new water supply project that would link the city with Spavinaw Lake more than 70 miles away. Displayed with the model were two of the huge water conduits that would bring water to Tulsa's citizens.[60]

As in the first show, the exposition emphasized continuity in the oil industry by spotlighting its history from the time of Drake's well to 1924. A specially arranged display of prints, drawings, and old photographs of oil men, oil equipment, and oil towns showed the spectrum of activity and development that had occurred in 65 years. Again Colonel Drake's tools were on display along with a full-size replica of his derrick.[61]

Contrasting with this exhibit was the display of the newest tools, supplies, and equipment, some of which, although fully operational and ready for distribution to oil operators, had intentionally been

A crowd waiting to see whose names were drawn for prizes from among those who registered at various exposition booths in 1924. This one was conducted by the Oil and Gas Journal. Courtesy Oil and Gas Journal.

kept off the market in order to be introduced at the IPE, the industry's "coming out party." The exposition allowed the showing to best advantage of such items as huge compressors, oil engines, gas engines, pulling machines, drilling machines, and pipe threaders before it was put to work in actual production.[62] "An oil show, in my opinion, helps to get new inventions, new ideas across to oil men quicker than any other way," comments Leslie Brooks.[63]

By the time the 1924 show ended, the IPE was firmly established as the premier oil show in the industry, and Tulsa's claim to the title, "Oil Capitol of the World," was stronger than at any time since it first was made in 1915.[64] All constituencies of the IPE obviously were increasingly attracted to the event: the worldwide oil industry, professional associations of the petroleum industry, and the City of Tulsa. In 1924 the total number of visitors was 27,000, almost double attendance in 1923, and the number of exhibitors was 86.[65] To accommodate these visitors, residents of Tulsa had opened their homes in record numbers.[66] Thus in all respects, the show in 1924 was a resounding success. Most exhibitors, visitors, and Tulsans agreed with the judgment of Paul H. Ehrhardt, a French mining engineer, who commented, "I hope that the exposition will be continued in Tulsa, the world's oil capitol, yearly so that oil men of the world can gather and exchange their ideas."[67]

Inevitably, however, there were some dissatisfactions both during and after the show. A few exhibitors said that the yearly format of the IPE was too burdensome because of the great expense of transporting their exhibits to Tulsa and staffing them once they were in place. However, the directors believed that the annual schedule was justified inasmuch as developments in the industry were occuring rapidly and needed to be presented to industry leaders as quickly as possible. They further reasoned that a yearly schedule was necessary, at least temporarily, to establish the IPE on a sound footing. Thus they voted to continue meeting annually and suggested that exhibitors who found the schedule inconvenient might choose to attend on alternate years.[68]

Another point of dissatisfaction among some participants and observers was the IPE's location. The most visible and vocal critic of the location on North Peoria was the *Tulsa Daily World*. The main objection seemed to be that the show was stifled in terms of future expansion so long as it remained at that location. It was true that in 1924 there had been congestion because of inadequate parking and narrow streets leading to and from the exposition grounds. Another consideration involved the cost of the Prairie Oil and Gas Company property. In his projected budget for the 1924 show, Manager McIntyre had estimated expenses at $121,900 and income at $181,500. This would have produced a healthy profit, but not included was $72,000 in taxes, rental fees, and fund guarantees required of the IPE Corporation. Thus when all accounts had been settled after the show of 1924, the IPE had a deficit of $13,000 for the year, a consideration which reinforced the contention of those who argued that the IPE should have its own exposition grounds.

The IPE directors did not allow themselves to become hotly engaged in the issue of the exposition's location. They knew that ultimately a new site would be needed if the IPE was to continue to grow as it had the previous two years. However, they believed that for the immediate future the facilities on North Peoria were more than adequate because these took the IPE "off the streets," giving it seven acres along with four brick and five steel structures and three structural steel arcades.[70] Moreover, a considerable expense had been incurred in decorating and beautifying that site. Between the indoor booths were columns, each column crowned with potted geraniums in bloom. The interior colors, green and white, were the official colors of the IPE. Grass, trees, and shrubbery had been planted throughout the grounds.[71] On the Peoria Street side of the grounds, two standard steel oil derricks had been constructed, and between them, at the Main and Brady Street entrance, hung an electric sign flashing "International Petroleum Exposition." Decorated archways were at every entrance, and the entrance on Admiral Place was specifically used by those who came by bus or foot.[72] All six entrances had gate turnstiles "of the latest and most modern type" purchased by the corporation at a cost of $720.25.[73] Because of these elaborate preparations at this site, the directors were not ready to make a hasty move—and thus they ignored the issue of the location.

1925

After the 1924 show closed, the directors of the IPE began work for the following year with new leadership and an unshakable dedication to continuing the exposition. Four members from the previous year remained on the board: Holden, McGraw, Hinderliter, and Heggem. The seven new board members included W. Merve Bovaird, secretary-treasurer of Bovaird Supply Company; Robert F. MacArthur, general manager of Barnsdall Oil Company; Oscar C. Staples, manager of the machinery department of Frick-Reid Supply Company (later Jones & Laughlin); D.D. Wertzberger, president of Wertzberger Derrick Company; Fred W. Insull, president of the People's Ice Company and Chickasha Gas and Electric Company; A.W. Leonard, vice president of Devonian Oil Company; and A.F. Bourne, secretary-treasurer of the Thompson and the Bradshaw Oil and Gas companies. As officers the board elected W.M. Bovaird president, Frank Hinderliter first vice president, Alf G. Heggem second vice president, William Holden secretary, and J.J. McGraw treasurer. Edward F. McIntyre continued as general manager.[74]

This roster of directors was not complete, however, for in February W.M. Bovaird submitted his resignation as president, citing conflicts of time between his duties with his supply firm and his work for the IPE. Two months later, R.F. MacArthur resigned from the executive committee, creating another opening at the highest level of control. Under normal circumstances these resignations might have impeded the work of the directors, but assuming the work was a man whose years with the IPE would become legendary, W.G. Skelly, founder and chairman of the board of Skelly Oil Company. He was named both to the executive committee and to the presidency.[75] Immediately he went to work to deal with the IPE's problems, principal among which were the dissatisfaction of some exhibitors with the yearly IPE schedule and the continuing controversy over a permanent site for the show.

Skelly contacted many of the large supply companies that objected to the yearly schedule and pointed out that rapid developments in the oil industry justified a yearly exposition at which buyers could examine and learn about the latest equipment. At the same time, he sent personal requests to his friends and associates in leading oil companies around the world asking them to voice their support for the 1925 show. He also asked the Mid-Continent Oil and Gas Association to urge its members to support the upcoming show.

Skelly tackled the problem of a site for the show

W.G. Skelly, president of the IPE from 1925 to 1957. Courtesy Oklahoma Heritage Association.

by throwing his support to an action already taken by the board. In March, a month before Skelly became president, the board had voted to buy the Prairie Oil and Gas Company property, which previously had been leased for the exposition. Skelly supported the concept of owning the site, believing that in the long run this would be cheaper and more satisfactory than a rental arrangement, and the $150,000 contract was signed. Although Prairie Oil and Gas previously had priced the property at $175,000, it agreed to contribute $25,000 of that to the IPE Corporation, indicating the spirit of the people by whom—and for whom—the IPE had been created.

There were other significant changes as the IPE prepared to celebrate its third birthday. After the 1924 show there had been complaints from those attending the Congress about the noise and distraction surrounding the exposition theater. To correct this, the directors separated the Congress from the exposition proper, arranging with the Mayo Hotel for Congress speakers to appear there. The Mayo was an ideal setting for the event; in 1925, the year the Mayo opened, it became the unofficial headquarters for exposition participants and distinguished visitors. The schedule of events was rearranged so that the events of the exposition would not conflict with the Congress, allowing everyone to participate in major events. Inasmuch as the Congress began at 10:00 a.m., the exposition gates did not open until noon; this allowed exhibitors to visit speakers without having to leave their booths. Another innovation enhancing the Congress was getting the Secretary of State of the United States to issue official invitations to attend the three-day Congress to every oil-producing nation in the world.[76]

President Skelly himself opened the 1925 conclave, expressing his hope that the oil industry could meet the challenge of a recent crisis in its image because of an inability of the public and the industry to understand one another. Skelly voiced his hope that there would be "better cooperation and a friendlier and more sympathetic contact between the oil industry and the public." The recent Teapot Dome scandal, he believed, had created an unfortunate misconception of the industry which all oil people needed to work to dispell. His theme was picked up by the next speaker, former Governor Henry J. Allen of Kansas, who asked that government officials present the public with a true picture of the oil industry. Other speakers also commented about the image problem. C.F. Kettering, president of General Motors, cautioned oil industry personnel to scrutinize their advertising to make certain that the highest standards of truth were observed.[77]

Other industry interests and concerns were presented by speakers such as Dr. Walter Scott, president of Northwestern University, whose topic was "Contributions of Universities Toward the Development of the Oil Industry." He represented a new contingent—education—included at the IPE for the first time in 1925. The American Petroleum Institute was represented by its president, J. Edgar Pew, who presented an optimistic picture of America's oil supply in his talk, "American Oil Resources—Demand and Supply." W.B. Guiberson spoke on "Taxation of Oil Properties," a topic of great concern to oil men, and he was followed by speakers representing South American oil interests.[78]

These changes probably saved the "Congress" part of the IPE from oblivion. Many participants in 1924 had been dissatisfied with the arrangements and schedule of the Congress, and they had recommended its discontinuance. Rejuvinated, it remained a part of the IPE.

There were other changes in the 1925 IPE as well as expansion. Because the entertainment element was highly popular, the directors in 1925 voted that this should be bigger and better. However, the plans for the Pageant of Princesses Parade, scheduled for the first day, went awry. Not only were 17 of the parade's most beautiful and intricate floats, representing $20,000, destroyed by fire on September 18, but also on opening day a torrential rain fell on Tulsa while the princesses waited to tour the downtown area. Officials reluctantly postponed the parade until the following day.

On Friday morning, parade officials and participants lined up at Eleventh Street and Cincinnati Avenue to lead Tulsans to the exposition grounds. With an escort of Tulsa motorcycle policemen and several local school bands, Grand Marshal Charles A. Holden, President Skelly, Manager McIntyre, city officials, officers of the Chamber of Commerce, 12 oil princesses, and King and King Petroleo rode in stately progress to the oil show. King Petroleo III was Frederick Ernest Windsor, son of a pioneer Pennsylvania oil man and former comrade-in-arms (during the Spanish American War) of W.G. Skelly and Edward F. McIntyre. Ramona Trees was Queen Petrolia for a second time because her successor, Virginia Burdick of Bradford, Pennsylvania, had not yet been chosen.[79]

The merchants and industrial parade, scheduled for Saturday morning, fared much better. It coursed toward the exposition grounds through crowds of excited Tulsans under a bright—dry—sky. It seemed that fate had singled out the merchants parade for favorable treatment, for the damp, gloomy weather that had halted the Pageant of Princesses returned on Sunday and plagued the exposition four of its remaining five days.[80]

Nevertheless, the show continued to the delight of distinguished visitors and participants. One group of visitors came en masse from Bradford, Pennsylvania, led by Grace Emery, daughter of the late Pennsylvania senator and manufacturer, Louis E. Emery.[81] She organized a delegation after a visit to the area the previous summer by IPE

Skelly Oil Company took a full-page ad in the Oil and Gas Journal to welcome visitors to the IPE in 1925, the first year of W.G. Skelly's presidency. Courtesy Oil and Gas Journal.

Manager Edward McIntyre and a group of Tulsans to promote the show. Securing Miss Emery's support was important because, thanks to her personal influence and that of her family name, a large group from Pennsylvania decided to attend.

The 150 Pennsylvanians traveled in six train coaches dubbed the "Emery Special." It was met at Tulsa's city limits by W.G. Skelly, J.J. McGraw, and Edward McIntyre, who boarded the train and escorted the visitors to a special platform erected where North Main Street intersected the tracks. A throng of Tulsans, many of them former Pennsylvanians, greeted the visitors as a local band played "Hail, Hail, the Gang's All Here." Each visitor's name was announced over a public address system as he or she stepped from the train, the last of whom was Miss Emery, who received a bouquet of red roses. A fleet of cars then transported the visitors to the Mayo Hotel, led all the way by the band playing "There'll Be a Hot Time in the Old Town Tonight." At the hotel there were further welcoming ceremonies, topped by a reception, a fashion show, a band concert, and a street dance.[82]

In 1925 the IPE was a true extravaganza. There were three parades, a carnival midway was added, and an all-new "Radio Revue" was booked for the exposition theater. Radio was the rage that year, bringing a revolution in entertainment, and among the "Revue" performers were several well-known radio stars from Chicago:

> Eddie Cavanaugh, story teller and character artist, was a star on Station WTAS. Miss Grace Wilson, a ballad and comedy singer, was a featured artist on Station WLS. Miss Dixie Fields, songstress and solo leader of chorus numbers, was from Chicago's Station WHT. Other performers included: Simmons and Clifford, the Melody Girls; Hal Gilles, dancer and comedian; the Rose Dress Four, the world's greatest roller skating act; Aaron Kids, a novelty kid act; and twelve Chicago chorus girls.[83]

Recreation also was provided by the second annual IPE Golf Tournament, an event which had proven extremely popular with oil men during the 1924 show. Again this was hosted at the Tulsa Country Club.

More was done to entertain and "show off" the 12 petroleum princesses who were competing for the title of Queen Petrolia. They were scheduled for luncheon dates with the Civitan Club, Junior Chamber of Commerce, Chamber of Commerce, Cooperative Club, and Lions Club. They were featured in exposition parades and appeared twice daily at the vaudeville theater, and there were formal receptions, balls, and a football game between the University of Tulsa and the Haskell Indians. The princesses were the focus of the glamorous social calendar of the 1925 IPE.

Social activity was less formal on the exposition grounds but strongly in evidence. Around exhibits, in booths, and in spontaneous "committees," oil people renewed old associations and made new ones. Much of this good fellowship occurred as a result of the work of the Knights of the Derrick, who met in the exposition theater on Thursday morning.[84]

The serious work of the IPE also was going forward, not all of it directly related to finding and producing oil. The first presentation of the IPE's first-aid meetings took place on the show grounds, an event sponsored by the American

W. C. Brown, Manager of Foxboro Headquarters at Tulsa, ready to give you instant airplane service

Aviation was becoming a tool of the oil industry by 1925. The Foxboro Company and its manager, W.C. Brown, offered air service in this ad. Courtesy Oil and Gas Journal.

Safety Council and the safety section of the Mid-Continent Oil and Gas Association. The winner of this competition received a silver loving cup. Teams from 24 companies competed for the prize, which was won by a team from Phillips Petroleum Company with a near-perfect score. Other teams placing high in the competition represented Mid-Continent Petroleum Company, Empire Refineries, and Skelly Oil Company.[85]

The real business of the IPE—showing new equipment and educating people about new techniques and procedures for finding and processing oil—was not slighted in 1925, although a few supply companies did not exhibit. They were missed, but "the show went on." There were new exhibitors and an increased emphasis on good fellowship. More space was given to arrangements conducive to relaxed conversation.[86]

An important consequence of the absence of several major supply companies was the opportunity accorded inventors to catch the attention of those attending. Inventors previously had been overshadowed by the grand and spectacular exhibits of the major suppliers, causing some inventors to lose faith in the IPE as a showcase for their work. Their success in 1925 restored their faith.[87] Another emphasis in 1925 was electrical devices intended for the oil field. Steam power had been the workhorse of the oil industry to that time, but electric power soon would establish its supremacy.[88]

There also were exhibits of new types of generators and procedures for making wire rope, along with other equipment and processes. Some exhibitors included hospitality accommodations at their booths; Marland Oil Company, for example, offered cigars, cigarettes, and pipe to-

The C & G Cooper Company, engine builders, took a serious attitude in its advertising for the 1925 IPE. Perhaps this was meant as a subtle comment on the "hoopla and ballyhoo" to which some exhibitors objected during the early years of the exposition. Courtesy Oil and Gas Journal.

bacco to those viewing its exhibit. Some firms offered musical performances or other types of entertainment.[89]

One popular exhibit was that of the Tulsa Radio and Electric Company, which kept IPE visitors up-to-date on the World Series between the Washington Senators and the Pittsburgh Pirates. General Electric Company attracted visitors with demonstrations of arc welding. The American Car and Foundary Company's display of tank car fittings also was popular, as was that of the Highland Body Manufacturing Company showing recent developments in truck cabs.[90]

As always, visitors were fascinated with scientific and technical exhibits, as well as with historic displays such as the memorabilia from the famous Drake Well and a replica of the derrick used in drilling that first commercial well in 1859. Shales, charts, and geological formations were shown along with a collection of priceless photographs of early-day wells. There also was a display of crude oil from various field around the world, allowing visitors to compare the densities and shadings of the many varieties of unprocessed oil. The lecture series of the Oklahoma and United States Geological Survey was popular, while Oklahoma oil men in particular were interested in a display of the geological formations of Oklahoma; large crowds could be found around this display at almost any hour the exposition was open.[91]

The mood was mixed as the 1925 International Petroleum and Congress closed. Fire and rain had hampered it, as had an unfortunate incident which occurred during the football game between Tulsa's Golden Hurricane and Haskell's Indians; after an opening ceremony during which the players received the ball from King Petroleo, a section of temporary bleachers, constructed to handle the overflow crowd, had collapsed as fans stomped and cheared, injuring 10 so severely they were hospitalized and slightly injuring another 400 to 500. On the positive side, the IPE's rain insurance paid $6000, reducing the $10,500 shortfall in expected receipts, and the corporation was allowed to postpone its payment to the Prairie Oil and Gas Company for the exposition grounds.[92]

In assessing the extent to which the exposition had become involved in pageantry and entertainment, President Skelly and Manager McIntyre reluctantly concluded that retrenchment was necessary. The exposition theater, they agreed, would have to be torn down and the spectacular revues and shows it had housed discontinued. These features had been highly popular, but they had cost more than they had contributed—and were, after all, incidental to the "scientific and educational" purposes of the IPE.[93]

In the really important areas, the 1925 exposition had been an unqualified success. Attendance was more than 11,000 over the 1924 total, and exhibitors had increased from 27 to 102 since the first show despite a number of supply firms' boycotting of the show. Further, exhibitors were reporting significant increases in sales with each show; in 1924 exhibitors had reported sales of $2,005,970 as a direct result of participating in the IPE.[94]

The IPE no longer was a fledgling, as it proved by setting new records in the face of its most trying year. It had a history showing a pattern of growth and an ability to adjust to the changing conditions of the oil industry. According to Don Barnum, "The shows had become an indicator of the industry, a measurer of the pulse beat. In 1923 the big attraction had been core drilling; in 1924 heavy machinery and new drilling methods had taken the spotlight, and in 1925 the emphasis had been on inventions and electrical devices."[95] And the hometown of the IPE was keeping pace, as Barnum noted:

> Just as the Exposition was nearing the point of going "over the top" in self-perpetuation, Tulsa too was beginning to get "over the top." The city had always gone out and sought companies and oil people to establish here, but now these companies were seeking Tulsa. By late 1925, with addition of the Transcontinental Oil Company and Independent Oil Company, the total number of oil company employees in Tulsa numbered 700. The city and its child, the IPE, were growing.[96]

1927

Having experienced losses owing to rain during the first three shows, the IPE directors took special care to select a date for the 1927 effort that would be least likely to see rain. All available weather records were reviewed, and these were correlated with almanacs to find a block of days that reasonably might be expected to be dry. All the evidence indicated that September 24 through October 1 showed the best prospects, and those were the days selected for the 1927 show. So certain were the directors that their meticulous research had found dry days that they chose to forego rain insurance for that show.

Unfortunately it began to rain on September 23 and did not stop until September 28. The directors could only unfurl their umbrellas and reflect on Edmund Burke's reminder to a colleague in 1791: "You can never plan the future by the past." They also watched dump trucks bringing crushed rock to spread over the dirt roads running through the new home of the IPE at the Tulsa Fairgrounds. J. Burr Gibbons, in his first year as manager of the show, summed up the battle against the mud in 1927, "We had forty-eight trucks hauling rock from five crushers for two days and nights in an effort to get out of the mud. There are places where crushed rock must be five feet deep on the grounds. It was disheartening to see yards of rock dumped and lost in the mud."[97] However, the rain

Buildings housing outside exhibits in 1927. The two-tiered structure in the upper left is the Texas Building. Note the muddy roads in which visitors and exhibits frequently were mired during the first five days of the show. Courtesy Oil and Gas Journal.

was not as disheartening as it had been in previous years despite the lack of rain insurance. The IPE in 1927 was stronger, headed in a new direction, in a new location, with a new manager, and under a new schedule.

The nagging question of how often the IPE should meet had been settled with the decision to cancel the show for 1926. President Skelly had pointed out to the directors that the debts of the corporation and the indifference of certain exhibitors strongly mitigated against a show in 1926. Two-year scheduling thus was forced on the exposition. At the same meeting at which the 1926 show had been canceled, a representative of the Tulsa Free Fair Board made an offer which provided a solution to the problem of the location of future shows. He told the directors that the Free Fair Board would like to join with the IPE and hold the two shows together. Under such a plan, the fair authorities would provide the IPE with exhibition space at the fairgrounds. The IPE directors found this offer most appealing, but something had to be done about the existing contract between the IPE Corporation and Prairie Oil and Gas Company. Fortunately the Prairie Company, always cooperative and supportive of the IPE, agreed to release the IPE from its contract and to allow the IPE to remove its buildings to the fairgrounds. Another nagging problem thereby was solved.[98]

There remained, of course, the problem of the recalcitrant suppliers who had boycotted the 1925 show. As members of the oil fraternity, their participation was needed, and the directors wished to reestablish and maintain unity within the fraternity. Therefore the directors voted to enlarge their membership and to spread control of the IPE throughout the entire spectrum of the oil industry. This action was taken on April 20, 1927, and membership on the board of directors was expanded to 25.[99]

Leadership on the board was not affected by enlarging the membership. W.G. Skelly continued as president, Alf G. Heggem and Frank Hinderliter as vice presidents, William Holden as secretary, and J.J. McGraw as treasurer. However, Edward F. McIntyre became a casualty of a dramatically changed emphasis in the exposition.[100]

McIntyre had been hired to manage the first exposition because of his experience with agricultural shows in Kansas and because of his connection with various carnivals. In 1927, however, the IPE directors decided to get out of show business. Such entertainment, they decided, could best be provided by the Tulsa Free Fair, which would coincide with the IPE. Thus McIntyre's talents no longer were needed, and he was replaced as general manager by J. Burr Gibbons, an early-day reporter for the *Tulsa World* and a civic booster (who later would start his own advertising company).[101] This shift in focus did not mean that fanfare and spectacle would be totally removed from the exposition, but that future spectacles would be more directly associated with oil industry concerns. The revues, actors, pageants, parades, princesses, and kings were to be replaced by equally entertaining and fascinating demonstrations and events from the real world of oil. J. Burr Gibbons' philosophy was that it was time for the IPE to get serious.

Nevertheless, there still was gaiety. From a band platform in the middle of the exposition complex, stirring live music was heard each day from 2:30 to 5:30 p.m., and in the evening there were IPE-sponsored dances on the grounds. Another feature that continued was the annual Petroleum Exposition Golf Tournament, held at the rain-soaked Tulsa Country Club. The winner in 1927 was Walter Critchlow of Ardmore, Oklahoma.[102]

Also dampened somewhat in 1927 was the Congress part of the exposition. President Skelly opened the Congress, speaking to a small group of suppliers, manufacturers, and visitors. In his

Suppliers, manufacturers, and exhibitors listening to President Skelly's opening address at the 1927 IPE. Courtesy Oil and Gas Journal.

keynote address, he recalled and highlighted the rapid developments which had occurred in the oil industry during the past three years. Also speaking at the 1927 Congress was its only foreign delegate, Manuel J. Zevada of the Department of Commerce and Mines of Mexico. After his talk there were comments from a few other people, and the one-day Congress then adjourned.[103]

The exposition opened on a grandiose note at 2:00 p.m. on September 24. President Calvin Coolidge was on hand to set off a "gusher," ample testimony to the stature the IPE had attained not only in its own industry but also in the nation as a whole. This "gusher" had been specially built, primed, and filled with oil for the occasion—a replica of Colonel Drake's derrick placed over a tank filled with oil from Drake's farm. A steam line ran from the tank to the top of the derrick and was rigged so that an electrical impulse would cause steam to flow through the line and force oil to spew from the top of the derrick. At 2:00 p.m. President Coolidge pressed a gold telegraph key, and the gusher did begin—indeed, "she blew." Too much steam had built up before the valve was opened, and when Coolidge pressed the gold key the spouting oil blew several boards off the top of the derrick, treating bystanders to a thrill usually reserved only for oil men. Following this "gusher," several oil men honored the old oil field tradition of dipping cattail weeds into the oil and lighting them to celebrate the completion of a successful well.[104]

The Mid-Continent Oil and Gas Association had planned this spectacular opening for the IPE, and the show's directors had cooperated enthusiastically. After it blew, the invocation was delivered by Dr. C.W. Kerr and the Akdar Shriners' band played "America." Then came a welcoming speech from Tulsa Mayor Herman F. Newblock followed by a speech by Oklahoma Governor Henry S. Johnston, who reviewed the success of the oil industry in finding a multitude of uses for petroleum. Then President Skelly rose to comment on the development of the IPE as an unparalleled showcase for the oil industry at which the latest developments in all facets of the industry could be spotlighted for the worldwide oil community. These opening ceremonies were colorful and tasteful and set the atmosphere for the rest of the show.[105]

An increasingly important aspect of the oil industry was safety. This had been acknowledged in the show of 1924 by a contest involving first-aid that had been of great interest to participants and visitors. Thus in 1927 it was repeated, the Mid-Continent Oil and Gas Association sponsoring the contest. Teams of six men each represented various companies in demonstrating the speed and efficiency with which they could treat oil field emergencies such as profuse bleeding and shock; they applied bandages and splints and gave artificial respiration to mock casualties to show how quick, proper first-aid could minimize the injuries of oil field workers. The prizes in this competition included a silver loving cup, a $20 gold piece for each member of the winning team, and a first-aid kit. Winning in 1927 was a team from the Empire Oil and Refining Company of Cushing, Oklahoma, and a team from the Marland Oil Company of Ponca City, Oklahoma, took second place. Tied for third were Sinclair Oil and Gas Company and the Skelly Company.[106]

Another feature from the show of 1924 was the "special day," greatly expanded in 1927. Each day during the exposition a salute was given to some town, state, or association; also singled out for recognition were certain branches of the oil industry and certain civic organizations.[107] A new feature in 1927, one in keeping with the emphasis on oil field safety, was a fire-fighting demonstration accompanied by a display of fire-fighting equipment and talks on safety procedures. Both the equipment and personnel for this event were supplied by the safety departments of major oil companies and the Federal Bureau of Mines. This particular event concluded with the fighting of an actual fire set in a pool of crude oil. Within 36 seconds the fire was completely extinguished by the application of a special foam created for the purpose.[108]

In 1927 the IPE started what became a tradition: awards to industry "Old Timers." For some time the show's management had been accumulating the records of long-time oil men who had served with distinction, and several types of awards were given. One was a gold medal to a "Grand Old Man" who had given at least 50 years of outstanding service. A silver medal, called the "Pioneer of Pioneers," was presented to the oldest living oil man with the greatest number of years of service. And bronze "Distinguished Service" medals were awarded to other contestants. The medals for these awards were donated by John D. Rockefeller, Sr., each inscribed on one side with a relief image of Rockefeller with an oil derrick in the background. On the reverse of the medal was the location of the IPE and the date.[109]

In 1927 there were 37 candidates for the gold, silver, and bronze medals. Several of those entered nominated themselves and wrote their own supporting letters; others were entered by friends. The winner of the "Grand Old Man" medal was J. Alexander Stephenson of Tulsa, an 81-year-old oil man who had been in the business for 67 years. Jacob Sheasley, age 93, who received the "Pioneer of Pioneers" award, was unable to attend the exposition, so the award was presented in his hometown of Franklin, Pennsylvania. Each award was announced by John D. Rockefeller during a banquet in the Governor's Suite at the Mayo Hotel.[110]

Friday, September 30, was special because Charles A. Lindbergh arrived under the sponsorship of the IPE, the Chamber of Commerce, the Mid-Continent Oil and Gas Association, and the local chapter of the National Aeronautical Association. His visit came less than four months after his heroic solo flight across the Atlantic, which had made him the most sought-after public attraction in the world. Thus his appearance at the IPE was considered a coup for the oil show's management.[111]

Lindbergh arrived at noon in his "Spirit of St. Louis" from Oklahoma City, flying into Tulsa over old Reservoir Hill, the downtown area, and the exposition grounds to land at McIntyre Airport (five miles east of downtown on Federal Drive). There he was met by Art Goebel, another world famous pilot who recently had flown from San Francisco to Honolulu in his plane, "Woolaroc," named for the home of oil man Frank Phillips, who had sponsored the flight.[112]

Lindbergh, Goebel, and a crowd of dignitaries made their way from McIntyre Airport to the exposition in a parade of cars. At the exposition there was a brief ceremony of welcome, after which Lindbergh was escorted to the Mayo Hotel and given time to rest before appearing as guest speaker at an evening banquet in the Crystal Ballroom.[113] On Saturday morning 2500 Tulsans turned out to bid Lindbergh and Goebel farewell as they flew over the exposition grounds in a final salute. Perhaps it was Lindbergh's visit that inspired Tulsa University's Golden Hurricane; on the day of his visit, the University's football heroes defeated the team from Parsons College at Fairfield, Iowa, by three touchdowns to one in what was billed as the first IPE football game.[114]

A main theme of the 1927 exposition, one in keeping with Lindbergh's visit, was aviation as part of the oil industry. The other themes that year were invention and science. Oil firms were invited to bring their airplanes to this exposition, and many of them did. Stanolind Oil Company stole this portion of the show by exhibiting its "giant" monoplane that could carry eight passengers and two pilots and fly at 110 miles per hour. The craft also featured sleeping berths, large observation windows, a washroom, a toilet, and a baggage compartment.

Other technological advances introduced at the show in 1927 were the seismograph and the torsion balance; both reflected the growing importance of geology in the search for oil and were viewed with fascination by IPE visitors.[115] The directors of the IPE understood that the technical and scientific side of the oil industry was where growth would occur in the future, so prior to the show of 1927 they decided to construct a separate Scientific and Technical Building for future use, a decision they made when all available space for such exhibits was sold out and there was still demand for more. Outside this building was an exhibit too large to be brought inside: a General Motors two-ton oil truck which, driven by the famous "Cannon Ball" Baker, had made a record trip from New York City to San Francisco in 137 hours, 36 minutes.

The scientific and technical section of the IPE also included a collection of oil field tools dating from the Drake well, a replica of that well, and an

Actual drilling of Chicken Farm Well No. 4 on the IPE grounds in 1927. The area under the rig floor was excavated so visitors could see the drilling in progress. Courtesy Oil and Gas Journal.

In 1927 the Hinderliter Tool Company's display of "honest tools" included tools used in drilling a well in West Virginia in 1859. Courtesy Oil and Gas Journal.

A view of the Scientific and Technical Building in 1927.

old "spring pole" oil rig; these were displayed near two 84-foot standard rigs. And, as always, there was an official IPE well in actual operation, the Chicken Farm No. 4. The area under the rig floor had been excavated so visitors could see the entire drilling process completely. Surrounding these displays and stretching across the exposition grounds was a dizzying variety of equipment and supplies used in the oil fields.[116]

Every factor worked in 1927 to put the IPE once again on a firm footing—except for the weather. It had been the best guess of the directors not to buy rain insurance, but September's rainfall total in Tulsa was more than double the normal amount. The result was that attendance was less that it otherwise would have been, and both the IPE and the Free Fair added an extra day and stayed open on Sunday, October 2.[117] Despite the rain, the IPE of 1927 showed a gain in attendance over 1925, and the number of exhibitors was up. Nearly 68,000 pushed through the turnstiles, and 189 exhibitors—almost 100 more than in 1925—rented space.[118]

So successful was the event in 1927 that by its end the directors were talking of reestablishing the event on an annual basis. J. Burr Gibbons told President Skelly on October 4 that "reports from exhibitors, oil company executives and exposition visitors indicate a greater variety of equipment and materials on display than in the 1925 show and the most complete exposition ever held."[119] Thus the fourth International Petroleum Exposition and Congress ended on a note of high optimism.

Contributing to this feeling was the IPE's substantially improved debt picture. After all accounts were paid for the 1925 and 1927 shows, as well as for the improvements made at the fairgrounds location, there was a cumulative debt of only $5000.[120] The fourth show had been a moneymaker. Finding a new home had worked wonders for the IPE—a permanent home that allowed space for it to grow to its full potential. The IPE had begun a period of growth that would last more than 50 years.

1928

At approximately 2:30 on the afternoon of October 20, 1928, Charles M. Schwab, chairman of the board of Bethlehem Steel Corporation, ended his opening remarks, lowered his uplifted arm, and set off a cannon volley that opened the fifth International Petroleum Exposition and Congress. As the cannon roared, a squadron of airplanes flew overhead, and the Oklahoma Military Academy band played "The Star Spangled Banner."[121] At the end of this stirring ceremony, pumps, drills, and refinery motors began to hum

Construction of exhibit booths and support platforms for the 1928 show. Courtesy Oil and Gas Journal.

The sidewalks surrounding the Texas Building, shown here in 1928, were constructed the year before because of the heavy rains that hampered the IPE in 1927. Courtesy Oil and Gas Journal.

and pulsate on the IPE grounds, while loudspeakers came alive with music piped from the bandstand in the center of the exposition. Later these same speakers would carry broadcasts by radio station KVOO from various points of interest on the grounds. The opening of the fifth IPE was both spectacular and stirring.

All these events had their origin months earlier when the directors met at the home of W.G. Skelly, president of the IPE, and voted that the show again would be yearly. They also decided, on the recommendation of Manager Gibbons, to hold the show later in the fall in order to miss the equinox—and, hopefully, the fall rains that had plagued previous shows. The dates selected were October 20-29.[122] Continuing as the officers were Skelly as president, Hinderliter and Heggem as vice presidents, and Holden as secretary; A.W. Leonard was elected treasurer to replace J.J. McGraw, who had died.

Going to work with unprecedented optimism, the board noted that the IPE had weathered its formative years and had made a permanent place for itself in its home city and in the worldwide oil community. There had been numerous improvements in the show's facilities at the fairgrounds, most of them at the suggestion of Gibbons. New concrete sidewalks had been laid to keep visitors from having to walk in mud as they moved from building to building. The Science and Technology Building had been enlarged; the entire area had been completely rewired; and the number of public conveniences had been increased. Also, the city had cooperated in providing better transportation to and from the fairgrounds; a new office had been built for the IPE staff; and the Oklahoma Building had been enlarged. By approving these improvements, the directors were voicing their faith in the future of the IPE.

Manager Gibbons capitalized on the expansion and improvements by an intensive publicity campaign, apparently believing that in past shows not enough had been done to "get the word out" to the people:

> In preparation for the 1928 show [Gibbons' office] disseminated 66,194 pieces of mimeographed "copy" for 350 newspapers and twenty journals. Approximately 300 separate articles were written for newspaper illustration, and other magazine stories brought the total of special "writeups" to about 450. Connections were also made with Associated Press, United Press, International News Service and the Consolidated Press news services.[123]

Gibbons' purpose was to orchestrate a multi-faceted publicity effort that would reach all the constituencies served by the IPE.

One of these constituencies was the foreign oil industry, for the directors were well aware of the importance of developing the international element of the exposition. Fortunately foreign oil men were interested, as were American oil men in displays brought from other countries. Letters, calls, and all types of contact were increased in 1928 to bring the international oil community into great prominence at the IPE while simultaneously not slighting the domestic industry. The IPE already had attracted the interest and cooperation of the American Petroleum Institute, giving it free exhibit space in 1927. In 1928 the IPE's identification with the API was even stronger.[124]

Again in 1928 that part of the IPE known as the Congress was limited to one day and was held at the Mayo Hotel, an arrangement which Man-

A main street of the show in 1928. Note the IPE patrol wagon, right, parked in front of the Police Department and Service Office. Courtesy Oil and Gas Journal.

ager Gibbon came to believe was detrimental to it. The Congress convened at 9:00 a.m. on October 20 under the chairmanship of D.W. Moffitt, vice president of Mid-Continent Oil and Gas Company. Following introductory remarks by Tulsa Mayor Don Patton, President Skelly introduced keynote speaker Charles W. Schwab, chairman of the board of Bethlehem Steel Corporation. Several other speakers followed representing the foreign oil community; they reviewed conditions in the industry outside the United States.[125]

The Congress concluded two hours before the official opening ceremonies took place so participants and visitors could reach the fairgrounds in time for that event. At 2:00 p.m. IPE officials, exhibitors, and visitors assembled before a reviewing stand surrounded by 43 flags representing the oil-producing nations of the world. On the platform were President Skelly, the Rev. C.W. Kerr, Mayor Patton, and Harry H. Rogers, president of the Exchange National Bank, who was to introduce the main speaker, Charles Schwab. Inasmuch as he already had spoken about cooperation that morning at the Congress, Schwab reemphasized his theme, reminding the audience that individual success was possible only in a climate of cooperation among all elements of the industry. When he completed his talk, Schwab initiated the cannon fire and other spectacular events which heralded the official opening of the IPE.[126]

The IPE would not have seemed official without rain, and it came on Sunday, the day a number of groups came from various Oklahoma towns to tour the exposition. Tulsa's weather was at its worst, but fortunately, after it blustered, blew, and threatened Sunday evening and night, the sky was bright and clear on Monday, and the weather never threatened the remainder of the exposition.

In 1928 the rearrangement of the IPE's priorities, made the year before, along with Manager Gibbons' campaign of publicity and the emphasis on the international aspect of the show, paid off handsomely with endorsements and support from several sources. The major oil related associations and groups were enthusiastic supporters, as were various state groups dealing with the industry; leaders who previously had taken a wait-and-see attitude were willing participants in 1928. The IPE was riding a wave of momentum.[127] Particularly successful was the international part of the show, for 20 foreign nations were represented: Canada, Colombia, Japan, Venezuela, Rumania, Austria, Denmark, Italy, France, Russia, England, Peru, Germany, Switzerland, Guatamala, Eduador, Argentina, Poland, the British West Indies, and Mexico. The 71 delegates from these nations were sumptuously entertained by IPE officials and local oil men.[128]

On the domestic side, the IPE likewise was expanding. The "Old Timers" event, which had attracted 37 registrants in 1927, drew 42 pioneers in 1928; they came together to form their own distinct group, known as "The Pioneers of the Oil Industry." On October 23, James Amm of New York was named "Grand Old Man of the Oil Industry" and received the Rockefeller medal, while Andrew Jackson Sanders, a Pennsylvanian who had followed the oil industry to Claremore, Oklahoma, was named "Pioneer of Pioneers" and received a silver Rockefeller medal.[129]

On Tuesday evening the "Pioneers" attended a banquet in their honor at which they were comended in several speeches for the 2208 years they collectively had given to the oil industry. "Pioneer of Pioneers" Andrew Sanders, 92 years old, arose to speak about the changes he had seen and was warmly applauded by those in attendance; they saw him as a symbol of the vigor and determined character of those who drilled into the earth for its riches. And Barney E. Harrigan, coordinator of the Old Timers program, was warmly thanked for making the event so successful.[130]

The main emphasis of the IPE, however, was on satisfying the nation's thirst for energy, and the star of the show in 1928 was rotary drilling.[131] Ever deeper wells were needed to find the elusive hydrocarbons, and the older method of drilling, cable tool, no longer was equal to the task. As the industry adjusted to the new method of drilling, the IPE helped show the way.

Another modern tool of the industry which caught the spotlight in 1928 was aviation. Air travel had become important to oil men in two ways: airplanes used petroleum products, and they were of value in helping oil men cover vast expanses of territory rapidly.[132] Visitors in 1928 flocked to aviation exhibits at the show not only because of the novelty of the machines but also because of the glowing predictions being made about the aviation industry. In addition to a display of single- and bi-winged aircraft, IPE visitors were treated to a demonstration of precision flying by a squadron of 20 airplanes.[133]

Evident at the IPE in 1928 was the expansion of technology as exemplified by $40,000 worth of new technical equipment on display in the Science and Technical Building, which had been enlarged that year. Complementing this equipment was a new and enlarged geological display, which was considered by experts to be the finest, most complete such display ever presented in the United States.[134] Alf Heggem, IPE vice president, by all accounts had outdone himself in directing and coordinating this segment of the show. And outside there were exhibits that included tractors, trucks, ditchers, and field repair cars.[135] Other large exhibits included a complete operating model of a refinery and an absorption plant showing how gasoline was made. Other displays included pumping units, storage plants, and compressors.

The exposition's facilities were ample for the show in 1928, and the directors could take pride that they had ample exhibit space, special features facilities, parking, and transportation—in short, all the components needed for a major industrial show. The physical facilities included three main buildings: the Texas, the Oklahoma, and the Science and Technical. These were arranged in a semi-circle around a courtyard expanse covered with chat, and there were paved roads and pathways in the area. This outdoor exhibit area in 1928 pulsated with a thunderous operation of heavy oil field equipment, including 12 steel and wood rigs, some actually making hole. Interspersed among the rigs were displays of the huge engines that powered the search for oil, some of them steam, others gasoline, and a few electrical.[136]

In 1928 it seemed that everyone in the oil industry decided to participate in the IPE, which was becoming a "family reunion" for oil men, with the result that attendance records were set. Paid admissions rose to 89,216, and there were 258 exhibitors and $10 million worth of equipment. More important, $3.5 million in sales were made by exhibitors during the exposition.

Perhaps even more important, Manager Gibbons could report after the 1928 IPE closed, that the exposition had become financially independent. In fact, a small surplus of nearly $5000 had accumulated in the treasury. Gibbons could say with pride, "I feel our show has been a great benefit to the petroleum industry, to Tulsa and to Oklahoma. Financially, it has been a complete success and I am positive its permanency for all time has been assured."[137]

Most IPE officials credited Gibbons' management with contributing significantly to the success of the show, and in 1928 the directors looked forward to many more advances under his leadership. But this was not to be. Undertaking the management of the IPE had cut heavily into Gibbons' time for operating his own business, and, after two years with the oil show, he felt compelled to concentrate on his own advertising company. He was highly pleased to submit a positive last report.

One portion of his report concerned a questionnaire distributed to exhibitors during the show which asked their preference concerning the frequency of the IPE, and Gibbons was delighted to report that a great majority wished to retain a one-year frequency. Another positive fact in the report was that almost half of the exhibitors in the 1928 show had attended for the first time. The directors found this fact particularly heartening inasmuch as it was proof that the drawing power and stature of the IPE were increasing dramatically.[139]

Gibbons did not leave the impression that no improvements could be made. He pointed out that in spite of recent additions of public conveniences, more facilities were needed. He reported that rest, refreshment, and telephone facilities were strained during the show. Another problem was the distraction created by the rumbling and roaring of the big engines in the courtyard exhibits. A number of exhibitors had registered strong complaints that their conversations with prospective customers were rendered unintelligible by the noise.

This demonstration of springpole drilling shows how slow and tedious this work was in the late 19th century. Courtesy Oil and Gas Journal.

Gibbons added to his recommendations that more emphasis be placed on attracting foreign visitors and a greater effort be made to persuade these visitors to bring exhibits. The goal of the IPE, Gibbons believed, should be to present exhibits from every oil-producing country. He pointed out that help in achieving this goal was available from federal agencies and other interests hoping to promote stronger relations with foreign nations.[140]

Continuing with his report, Gibbons expressed some frustration at being unable to attract greater interest in the IPE on the part of the petroleum marketers. This contingent of the oil industry had for many years conducted its own industrial shows and had shown great reluctance to abandon them in favor of joining the IPE. Gibbons noted minimal success with this group and said he had experienced considerable difficulty in getting marketing and filling station displays. He concluded that the effort to involve the marketers should be continued.

Gibbons also was dissatisfied with the current arrangements for the Congress which he believed suffered because it was separated from the rest of the exposition. He thought an auditorium should be built behind the Science and Technical Building to house the Congress and similar events. He also believed that the Congress session should be expanded to one full day or divided into several short sessions over several days. The Congress, he reasoned, had more enduring news value than other events since "the exhibits, once described, do not change and attendance from day to day is made possible by the space the show can secure in the press."[141]

In this connection, Gibbons recommended that the IPE seek out more events of the forum type, such as association meetings, councils, conventions, and conferences. He pointed out that newspapers were interested in reporting the proceedings of meetings, and these reports would keep the exposition before the public. Attendees at these meetings would also increase attendance at the show, he said.[142]

With this thorough report, Gibbons concluded his official duties for the IPE, and the show itself closed down at 9:30 p.m. on October 29. Groups of Indian dancers gathered in the open plaza at the center of the grounds and began to dance to the music of tribal drums as the heavy engines on display wound down and stopped. The costumed Indians formed a snake dance and led a long troupe of IPE officials, participants, and visitors to the main gate of the exposition grounds where the entire group stopped and fell silent, waiting for the last official sound of the 1928 IPE. It came at 10:00 p.m.—piercing whistle blasts from four of the drilling rigs which had been at work in the courtyard exhibit area throughout the exposition. When these blasts trailed off into silence, the fifth International Petroleum Exposition and Congress was ended.

1929

Searchlights swept the sky and anti-aircraft guns thundered, rhythmically filling the sky with brilliant shafts of fire which leaped toward a screaming bomber in the midst of a savage attack on the throngs frozen in amazement on the grounds of the sixth International Petroleum Exposition and Congress. This mock raid and staged display of anti-aircraft bombardment was a nightly feature of the 1929 IPE and underscored the emphasis placed on aviation during the show. In addition to the display of aerial warfare, closed-course air races were featured on Monday, Tuesday, Wednesday, and Thursday afternoons. These races were of two types: civilian free-for-all races, and competitive flights among army and navy aircraft featuring pursuit, attack, and observation aircraft. The races began at the Tulsa Municipal Airport, proceeded to the fairgrounds where the planes touched down, continued on to the Garland Airport one mile west of the KVOO radio tower, and concluded when the planes crossed the finish line at the fairgrounds.[143]

President Skelly, a strong advocate for boosting aviation at the IPE, set aside Thursday night for the conclusion of the aerial events. A ceremony was held at which several trophies and almost $8,000 in cash prizes were awarded to the fliers who had won or placed in the various competitive events. Skelly believed that aviation and the oil industry would draw closer and closer as both industries developed and grew, and he predicted that aviation events would figure in all succeeding oil expositions.[144]

Not everyone, however, felt as comfortable as President Skelly with the marriage between aviation and the petroleum industry, at least not as far as the IPE was concerned. The show's new manager, W. B. Way, thought the aviation displays were too distracting from the real purpose of the exposition which was to feature, spotlight, and emphasize the oil industry. He agreed with Skelly that the aviation segments were enormously popular, but he argued that gate receipts were not commensurately increased by the attention given to aviation.[145] The IPE, however, was only one of the many public events featuring aircraft shows in these years. Developments in air travel had captured the public's fancy, and people who attended such events as the IPE expected to see some action in the air. Certainly no perceptible damage was done to the IPE, and the aircraft displays satisfied a strong curiosity of the public. Time, however, vindicated Way's judgment.

The directors' good judgment in selecting Way to succeed J. Burr Gibbons as general manager was confirmed. Way hailed from New Jersey where he began his professional life as a traveling salesman for a valve manufacturing firm. After proving himself to be an enthusiastic and committed member, he was elected secretary of the Natural

A crowd listening to the keynote address of E.B. Reeser, president of the American Petroleum Institute, in 1929.

Gas Supply Men's Association. Later he was chosen secretary of the Natural Gas Association, a position he held for 10 years before receiving a call from the IPE. Director A. W. Leonard strongly recommended Way after hearing many positive comments about him at a convention of the Natural Gas Association.[146]

Having secured a new manager in whom they had complete confidence, the directors moved ahead with plans for the 1929 exposition, electing to retain the same slate of officers. President Skelly optimistically pronounced a sentiment shared by all the officers and directors of the corporation: "This year witnesses the culmination of several years of planning and building the IPE to the point where it ranks with the great industrial shows of the world."[147]

Work was begun on expansion and improvement of the exposition's facilities. More conveniences were added for the comfort and refreshment of exposition visitors. The most significant addition to the facilities was a new auditorium constructed at the back of the Science and Technical Building and intended primarily to house the Congress and bring it into the mainstream of the exposition. This relocation did not signal an expansion of the Congress feature as it remained a one-day, three-hour event. In fact, the Congress was continuing to decline in importance among IPE events, but, according to President Skelly, its relocation to the exposition grounds marked the final step in unifying the IPE in its own quarters.[148] This step, he said, augured well for the IPE.

The official opening of the show occurred when

President Skelly and the special guests of the IPE ascended the speaker's platform to sound the show's keynote and to summarize achievements over the past year. President Skelly was accompanied by the Reverend C. W. Kerr and Tulsa Mayor Dan Patton, who welcomed guests and participants to the show and then surrendered the lectern to their distinguished guest, F. B. Reeser, president of the American Petroleum Institute. Reeser's presence signalled a new era of influence and prestige for the IPE, for it indicated acceptance and recognition by one of the oil industry's foremost professional organizations. In his address Reeser underscored the significance of this recognition, "A very significant factor in the 1929 exposition is that the API has joined hands with the directors of the exposition in making the show a success. It is of greatest importance to the oil industry that the two large organizations are working side by side in the exposition."[149]

The remarks by platform guests were followed by a stirring display to denote the conclusion of preliminary events and the beginning of the show proper. A small girl tugged on a rope to unfurl a 120-foot American flag, the world's largest, between two oil derricks, and a band struck up the National Anthem while aerial explosions unleashed tiny American flags from specially made fireworks bombs. For days visitors to the show could be seen carrying these miniature flags which they had caught from the air on opening day.[150]

All, however, was not tranquil in the oil industry in 1929, and the sixth International Petroleum Exposition and Congress reflected the dissatisfaction of oil men over the flood of oil that was coming to market and driving prices lower and ever lower. The 1927 and 1928 shows had also reflected this concern, but the situation had not improved, and the anxiety had continued to grow. What started it all was the 1926 discovery of the Greater Seminole Field, from which oil streamed in awesome quantities, and the situation was exacerbated in 1928 when the Oklahoma City Field was brought in.[151] "Overproduction" was heard everywhere, and oil men were locked in debate over the pros and cons of placing controls on the flow of oil.

Control had, in fact, been imposed shortly before the 1929 IPE opened, and the Oklahoma City Field was closed down for 30 days. At the exposition itself, the sentiment favoring proration was in the ascendancy and carried the day at a large gathering of Oklahoma oil men held on October 8 in the assembly room of the Exchange National Bank. The motion was made and approved that the assembly of 125 operators go on record as favoring a plan to reduce Oklahoma's daily production from 800,000 to 650,000 barrels.[152]

Another important issue in the minds of oil men in 1929 concerned standardization of oil field equipment, and when the American Petroleum Institute's Mid Continent District Division of Development and Production Engineering met in the exposition auditorium, Sun Oil Company's President, J. Edgar Pew, called for moderation. Their deliberations, he said, should be governed by "intelligent, collective and cooperative thought."[153]

Not all of the meetings in the new exposition auditorium were so solemn or so narrowly focused on oil industry business. The initiation of the new facility was marked by a wide variety of other events, one of these a petroleum industry film series featuring dramatizations of various phases in the discovery, production, and processing of oil. Films with titles such as "The Story of Gasoline," "The Story of Lubrication," "The Story of a Mexican Oil Gusher," and "The Story of a Rotary Drilled Oil Well" attracted throngs of viewers who were curious to know the inside story of oil. Nor were the activities held in the new auditorium exclusively secular. On Sunday, October 6, a cathedral atmosphere prevailed as Salvation Army members gathered to conduct a religious ceremony followed in the afternoon by a choir recital of sacred music.[154]

The mood shifted again on Wednesday, October 9, when the Pioneers of the Oil Industry Association convened in the auditorium for its first meeting as an incorporated body. It was Old Timers Day at the IPE, and the oil industry pioneers, 130 strong, met to celebrate their emergence as an "official presence" at the "World's Fair of the Oil Industry." From their ranks they selected 75-year-old John C. Looker as the "Grand Old Man of the Oil Industry" and 72-year-old Martin Moran, president of the Texas Pipeline Company, as "Grand Old Man of Tulsa."[155]

The good times rolled on at the IPE in spite of the temporary consternation that overtook the oil industry because of the unrelenting stream of crude oil flowing from Oklahoma's great fields. The sixth International Petroleum Golf Tour-

The gate of the 1929 IPE on 21st Street. The crowd was just moving forward to the exhibit area following the opening ceremonies. Courtesy Oil and Gas Journal.

nament, in progress at the Tulsa Country Club, was enjoying unprecedented participation on the part of delegates, officials, and exhibitors. Other social activities included dinner meetings, receptions, and cocktail parties.[156]

Good times also prevailed on the exhibit grounds and at the entry gates to the show, where more than 100,000 visitors viewed the displays of almost 300 exhibitors. The equipment and supplies they had witnessed were valued at approximately $12 million, another IPE record. The overproduction-proration annoyance that was currently troubling the industry and the economic disaster that lay ahead for the nation could not be detected among the enthusiastic throngs coursing through the exposition grounds. The IPE as a whole had reached new heights, especially in terms of the recognition it had been accorded by the prestigious American Petroleum Institute. The exposition had become the shining star of the oil industry.

Manager W. B. Way had joined the IPE in a fortuitous year, but he thought that further improvements were in order. In his final report to the directors about the 1929 show, he stated that the refining and marketing branches of the oil industry needed further enticement to secure their participation. He recommended that separate facilities be provided by the IPE Corporation to the marketing and refining companies. He believed that expansion, in general, was needed, and he called for enlargement of the Texas and Oklahoma buildings.

Way was even more adamant that something be done to improve the lodging situation for the visitors the IPE brought to Tulsa. He was indignant in the extreme at what he considered the exorbitant prices local hotel owners were charging visitors, and he wrote scorching letters to them. The thrust of his comments was that the hotel owners were strangling the "golden goose" by raising the guaranteed prices which the IPE housing committee had announced in pre-exposition publicity. The IPE, the City of Tulsa, and the hotel owners themselves would suffer, he said, unless IPE visitors could expect reasonable prices for their lodging.[158]

Way also was concerned about what he considered to be the undue emphasis still being given spectacle in the exposition, and he closed his report with these comments: "So if we devote all our effort to the show feature and lose sight of the fact that the oil company participation is really the most important feature, we are harking to the sound of the tin horn and closing our ears to the music of the band.[159]

1930

The sixth International Petroleum Exposition and Congress closed on a note of triumph, having enjoyed growth and expansion in every significant aspect. The economy and the nation in general, however, were facing upheaval and despair, and the bright economic prospects of America dulled. Within a few days after the close of the 1929 IPE came the Great Crash, dashing financial empires into useles shards and instilling fear and insecurity into the financial heart of the nation. The IPE directors decided that nothing was to be gained by acquiescing to the anxieties of the time and closing a successful enterprise. They believed it was their duty to reopen the gates of the IPE in 1930 as a signal to all that the oil industry would not take flight because of the wolf at the door.

When the show opened at 1:00 p.m. on October 4, those officiating tried to keep the tone optimistic. Robert P. Lamont, United States Secretary of Commerce, gave the keynote address in which he expressed a strong hope and belief that America soon would work its way out of its economic distress. He said that Oklahoma was making its contribution to recovery by continuing to do what it did best, which was to explore the plains and valleys of Oklahoma for the fuel the nation lived on. He reminded his audience that almost

Visitors inspecting exhibits on a main street in 1929. On the left is the Oil and Gas Journal booth. Courtesy Oil and Gas Journal.

30 percent of the United States' production of oil over the previous five years had come from Oklahoma, as well as 18 percent of the world's total. Nor did he overlook the current concerns about overproduction, a question with which oil men had been preoccupied for five or six years. Lamont explained that the Hoover administration wished to refrain from imposing a government solution on the problem. The oil industry itself could be most effective in meeting the challenge of overproduction, Lamont said, because the industry was keenly aware that hydrocarbon fuels had finite limits and that future generations would depend on the wisdom with which the present generation expended available reserves. The best thing the federal government could do was to provide any assistance which was requested by the industry, and he reported that the Hoover administration stood ready to respond when called upon. He also said that the Federal Oil Conservation Board was intensely studying the problem for the purpose of making helpful suggestions.[160]

Turning his attention to the country's economic woes, Lamont expressed confidence in the strength and durability of the American system. His main theme was "maintain confidence," and he assured his listeners that America's economy was, at base, as sound as ever. "Make sure of this," he concluded, "we shall emerge from this depression more efficient than we entered it…. Maintain confidence, we shall soon be on the upturn."[161]

Secretary of War Patrick J. Hurley repeated the optimistic theme but focused his comments on the local audience. "Don't sell Tulsa short," he said. "Don't sell Oklahoma short and don't sell the United States short."[162] A stirring moment followed the secretary's remarks as the IPE's giant version of Old Glory majestically unfurled, the strains of the national anthem filled the outdoor assembly area, and rockets burst high above the assembled throng, releasing showers of miniature American flags.[163]

On this same Saturday another significant ceremonial opening was taking place. Tulsa's famed Skelly Stadium, gridiron home of the University of Tulsa, was being dedicated. Participating in these ceremonies were Hurley, Robert P. Lamont, E.B. Reeser, Henry L. Doherty, and Tulsa Mayor George Watkins. Of course, Bill Skelly was prominently in attendance, having donated the $125,000 which set the wheels in motion to make the stadium a reality.[164]

In 1930 Skelly was in his fifth year as president of the IPE. The other officers also were veterans, leaving the executive committee unchanged for the third consecutive year. The show was a big winner and none of the executive committee members wished to break up their successful team. The health of the show was attested by the continual demand for more exhibit space, and the financial success of the 1929 show made possible a $100,000 expansion of the facilities in 1930. Two new buildings, the Marketers and Refiners Building and the California Building, had been completed during the summer of 1930.[165] Additions also were made to the restaurant facilities, and the decision was made at the urging of Way for the Exposition Corporation to assume operation of food service on the grounds rather than continue with lease arrangements with private individuals. Further expansion on the grounds had been undertaken by some of the major oil companies which were in the process of erecting their own exhibit facilities.

Improvements also were made in the transportation to and from the grounds. Day-long shuttle-car service under the direction of the Junior Chamber of Commerce was provided for the first time, and improved schedules were offered by the companies providing taxi, trolley, and bus transportation.[166] Better communications were available through the Western Union branch and the post office sub-station which were housed in the replica of the Drake Well, making it possible for oil executives to continue with their daily business operations while they enjoyed the exposition.[167]

Another major change announced at the 1930 show was cancellation of plans for a fall exposition in 1931. The directors had determined that the purposes of the exposition would best be served by shifting the opening of the next show to the spring of 1932. The spring date would coincide with the period of heaviest activity in the oil industry and would attract livelier interest on the part of the oil companies. It would also support the planned convention of the Natural Gas Division of the American Gas Association.[168]

The exposition directors had found that many oil industry associations and professional societies would arrange their meetings and conventions to coincide with the IPE if it were convenient, and the directors thought they should cooperate to the fullest extent possible. They believed that the activities of all branches of the industry should complement each other and engender unity within the ranks, and they were of the opinion that the IPE could be a strong force in achieving this goal.

At the 1930 show one of the events which strongly supported this perception was the convention of the prestigious American Society of Mechanical Engineers, which held twice-daily meetings during three days of the exposition. A topic of continual conversation during these meetings was the advances that had been made in the production of gasoline, primarily through the development of the cracking process. Lyman C. Huff, chief engineer of the Universal Oil Products Company of Chicago, said in a speech before the Society that he considered the cracking process to be the most significant development in refining in more than a decade. He stated this development had made possible the production of a superior grade of gasoline to keep pace with the higher performance of American cars. It had also made possible a dramatic increase in the volume of gasoline that could be extracted from given quantities of crude oil.[169]

Another meeting, one which had become a staple of the exposition, was the annual safety first-aid meet sponsored by the Mid-Continent Oil and Gas Association. An addition to this event in the 1930 show was the participation of the Boy Scout organization; it presented its own first-aid demonstration on October 11, the last Saturday of the show.[170]

One consequence of the increased participation by oil industry associations and professional societies appears to have been the termination of the Congress. It was not announced or held as a separate event during the 1930 show presumably because its function had been superseded by the separate meetings of the various industry organizations. It had served its purpose by showing the value to the industry of centralizing meetings and forums emphasizing oil industry developments, problems, and concerns. Change and readjustment are staples of the oil industry, and in the passing of the Congress the IPE was merely reflecting the dynamism of the industry it served.

But tradition was important to the industry, and beginning in 1927 it had taken time out dur-

The Quigley Company showed a variety of fluids used in the oil industry in its indoor display booth. Courtesy Oil and Gas Journal.

ing its "World's Fair" to honor its founders, the Old Timers. Again in 1930 special ceremonies were planned by the Pioneers Association. About 100 of these oil industry trailblazers met on October 8 in the exposition restaurant to enjoy lunch and to reminiscence. In ceremonies held later, they presented their honorees of the year: "Grand Old Man" E. H. Sloan of Pittsburgh, Pennsylvania; and "Pioneer of Pioneers" J. J. Larkin of the Larkin Torpedo Company of Tulsa.[171]

An unplanned event at the 1930 show was a protest parade consisting of several bands and nearly 300 cars which coursed through downtown Tulsa to publicly express the dissatisfaction of some oil men over the lack of federal protection for domestic oil production. This group believed that the economic problems of the American oil industry were directly related to the influx of foreign oil. Tariff protection was needed, they believed, to solve the problems of overproduction and proration.[172]

Despite the sentiments expressed by the paraders, foreign participation in the IPE was strong. Twenty foreign countries sent 80 delegates, and two countries, Poland and Venezuela, reserved booth space.[173] The evidence was never stronger that the IPE was the worldwide forum for the oil industry and was taken seriously by all branches—and factions—of the industry.

Seriousness of purpose was indeed becoming more and more evident in the planning of events and activities for the IPE. Non-industry related entertainments were increasingly subdued with each succeeding exposition, and in 1930 such activities were severely limited. Acrobats Roxey La Rose and Daredevil Gates appeared twice daily on the exposition grounds to entertain crowds with tight-rope walking and rig-climbing acts, but no

William B. Way, general manager of the IPE from 1925 to 1928, sits in his office in 1930. Courtesy Oil and Gas Journal.

troups of entertainers appeared as in the past. On opening day the Boys' Mounted Troup of America from Baxter Springs, Kansas, entertained visitors with demonstrations of horsemanship and roping. An air show also occurred during the exposition, but it was confined to the municipal airport and did not interfere with activities on the IPE grounds. Entertainment of another type, the seventh International Petroleum Exposition Golf Tournament, was held at the Tulsa Country Club and drew heavy participation.[174]

These diversions did not detract from the business of the IPE, yet they provided pleasant breaks along the way. The exposition had reached a high point in 1930 and settled into its role as the pace setter for the oil industry. Manager Way's report for 1930 was upbeat in every respect. The value of exhibits had risen to $13 million and the number of exhibitors to 321. Gate counts showed that 112,605 visitors had viewed the attractions and exhibits of the show, an increase over the previous year's attendance although some decline in gate receipts was noted. This decline, in fact, was taken as a positive sign by the directors that more oil men and fewer disinterested visitors were passing through the gates. Overall net profits amounted to $25,000 which would become the seed money for the next show. When all data were tabulated, the directors concluded that the IPE had shown a 30 percent growth over the previous show.[175]

Particularly pleasing to the marketers and refiners was the popularity of the exhibits they had presented in their new building. Not only had their exhibits attracted the curious but also the serious customer. Many business negotiations had taken place and a considerable volume of sales was tallied during the show. These exhibitors had found the IPE to their liking and most of them began immediately making plans for the projected 1932 show.[176]

That show, however, was never to be. The economic condition of the nation did not immediately take the turn predicted by War Secretary Hurley and Commerce Secretary Lamont in their opening speeches for the seventh exposition. The financial situation remained dark, and it was not until 1934 that the directors thought the prospects were bright enough to again open the gates on the great oil show.

CHAPTER TWO

AFTER THE DEPRESSION AND BEFORE THE WAR
(1934-1940)

The IPE marked time after the 1930 exposition until it opened again in 1934. In the interim American business, the petroleum industry included, wrestled with the Great Depression. When the reopening came, representatives of the federal government were on hand to express eager confidence in the IPE and what it represented in terms of the ideals of free enterprise. The IPE became a symbol of America's best hope for regaining its financial footing and retaining its political freedom.

The depression had sobered American industry somewhat, and the IPE reflected this in an increased seriousness of purpose. Displays of equipment and supplies were supplemented with the first presentation of the Hall of Science which became the educational arm of the IPE. Technological development in the petroleum industry had become explosive, and oilmen from every corner of the world had begun to rely on the IPE to update their knowledge so they could remain viable in a fiercely competitive industry and in a world preparing for war.

Sideshow-style entertainment and ballyhoo had not yet totally disappeared from the IPE—and to a certain degree they never did—but priorities for the great "World's Fair of the Oil Industry" were multiplying and more grumblings were heard that diversionary entertainments were not compatible with the IPE's mission as the petroleum industry's premiere showcase and forum. Many believed it was time for the prince and the showgirl to part company so that the prince could concentrate his energies on "tending to business." Entertainments became increasingly sedate and subdued and "business" moved into the spotlight. On December 8, 1941, the IPE itself stepped aside so that the industry it represented could tend to a deadly serious business.

1934

The diamond jubilee of Colonel Drake's founding of the oil industry occurred in 1934, and, fortunately, economic conditions in the industry had crept far enough out of the doldrums gripping the country that a show could be held. Had the anniversary fallen in 1931, 1932, or 1933, the case may have been far different, and there was much anxious discussion among the show's directors during the fall and winter of 1933 before the 1934 show was settled upon.[1] But when the decision was made, the directors stood behind it unanimously.

From its beginnings the industry had existed in an atmosphere of risk, so it was not surprising that its "World's Fair" would be resumed at the earliest possible moment. The decision was applauded by some of the most respected figures in the industry, as was indicated by a letter received by Manager Way from Harry F. Sinclair which declared:

> I learned with satisfaction that the International Petroleum Exposition is to be revived this year. While the skies are not all clear, much progress has been made toward re-establishing our industry on a sound basis. "Dollar oil" is certainly better than fifty-cent oil, and the outlook for further stabilization extending through the marketing division of the industry is good enough to encourage effort to make this year's exposition a celebration of progress toward recovery in the oil business.
>
> My best wishes to you for success in this effort.[2]

The IPE had been misssed, and its resumption promised to spark unprecedented enthusiasm and participation. Predictions were that the eighth exposition would be the "largest show of inventions, new equipment, and devices used in the oil industry" yet held.[3] Way commented that producers and refiners had undertaken plans to improve and extend their properties on the exposition grounds and to make significant additions to their equipment. He said 1934 would be the "opportune time" for exhibitors to display new machinery and inventions "which have been perfected by engineers and scientists over the past three years in which, due to depressed business conditions (they) have not as yet been generally introduced to the industry."[4] Interest was high, Way pointed out, because the IPE, "a non-profit institution owned by the oil industry" had gradually acquired a complete plant, and the executive committee had reduced rates for exhibitors by 10 percent for inside accommodations and 50 percent for outside spaces.[5]

As in other years, a well-organized program of publicity was set in motion for notifying the worldwide oil industry that its showcase would soon reopen after three years of inactivity. A joint resolution introduced in the United States Senate by Senator Thomas P. Gore called for the United States government again to endorse the IPE and to issue invitations to all states of the Union and to 50 oil-producing countries.[6] On the local scene a seven-step approach for publicizing the show was announced by President W. G. Skelly: more than 50 railroad and airline officials planned to invite more than 18,000 oil men from 38 states who attended the 1930 show; major marketing companies invited thousands of dealers and jobbers to attend; marketing associations in a dozen states sent out advertising material in their bulletins and in letters to their members; President Roosevelt signed the joint-resolution passed by congress, and the local IPE staff was submitting suggested lists of invitees; every exhibitor agreed to attach advertising stickers to letters they prepare to invite friends and customers to the show;

Aerial view of the IPE in 1934. The long structure in the foreground is the Texas Building; the long structure in the upper portion is the Oklahoma Building. Between them are buildings of individual exhibitors and the show's refinery building. Courtesy Leslie Brooks.

major companies were advertising the show in radio programs they sponsored, and Tulsa radio station KVOO was broadcasting weekly half-hour programs advertising the exposition, and the National Broadcasting Company planned to broadcast some of the programs to the entire country over its chain of stations.[7]

The Tulsa Chamber of Commerce also was mobilized to spread the word. Its petroleum exposition committee organized a group of speakers to appear at various public meetings in towns all over Oklahoma to give talks about the exposition. Each speaker was thoroughly briefed on the show so he could give an accurate overview of the IPE, answer questions, and encourage attendance on the part of the general public.[8]

The 1934 exposition also saw a continued increase in interest on the part of the oil industry's professional associations, more of which were scheduling meetings to coincide with the IPE. Three of the largest associations notified exposition officials as early as January of 1934 that professional meetings were being scheduled during the show. The power division of the American Society of Mechanical Engineers announced meetings, as did the American Association of Petroleum Geologists, the Mid Continent Oil and Gas Association, and the Missouri Oil Men's Association. Other groups which contacted Manager Way about possible meetings were the Iowa Petroleum Association, The Gasoline Division of the Texas Automotive Maintenance Association, the American Gas Association, and the Independent Petroleum Association of Texas.[9] Numerous other associations connected with the oil industry scheduled meetings during the week previous to the show so that their members could be in Tulsa to view the oil show exhibits.

This intense activity on the part of the professional societies gives a clue to the conspicuous absence in the literature of the time of releases, notices, and announcements concerning the Congress feature of the IPE. The *Oil and Gas Journal* noted, "Instead of general addresses, meetings of special interest have been scheduled by various organizations."[10] The Congress had served its function grandly when the concerns of the oil industry were capable of being generalized in a half-day program. The explosion of technology in the industry had inevitably brought about greater specialization and a need for numerous meetings with specialized concerns.

After all the meetings, speeches, announcements, publicity notices, and preparations, May 12 dawned somewhat blustery for the official beginning of the eighth International Petroleum Exposition and Congress. (The word "Congress" was retained in the title even though the feature had virtually disappeared.) One report said that the "wise" attendees "wore dark glasses, and that is a tip for visitors to the Exposition."[11] The official opening was given a unique twist by being signalled with the press of a button in Washington, D.C., by Axtel J. Byles, president of the American Petroleum Institute. Visitors were welcomed by Tulsa Mayor T.A. Penny, who declared:

> This occasion, so brilliant and vivid in its immediacy and so impressing in its scope, calls to mind the fact that within one generation a great oil indistry has come to its overshadowing emminence and the City of Tulsa has come from remoteness and humility to the status of international host. As a pioneer citizen of this community and Oklahoma who has witnessed these wondrous changes and as the present mayor of Tulsa, it is my proud privilege to welcome you availing people from everywhere.[12]

President Skelly added his welcome to the mayor's with these comments: "We have assembled here today to formally open the eighth International Petroleum Exposition and Congress, and as its president I take great pleasure in extending a cordial and hearty welcome to each and every one of you."[13]

The prelude to these speeches had been the invocation of Reverend George McDonald, pastor of the First Methodist Church who prayed "so fervently about the oil business and how oil men should seek divine guidance in directing their future policies that one petroleocrat remarked, 'That's the first petroliferous prayer I ever heard, and I for one like it and wish we could have more prayers for the future of the business.'"[14]

These ceremonies were followed by lighter rituals, one of which was a band concert given by the marching band of Oklahoma A & M College (now Oklahoma State University). The most colorful of the concluding rituals was the "March of Nations," performed by 50 Boy Scouts carrying the flags of the various countries represented at

Tulsa Mayor T.A. Penny welcoming visitors at the opening of the 1934 IPE. Courtesy Oil and Gas Journal.

Opening ceremonies for the ninth IPE were concluded by an assemblage of Boy Scouts carrying the flags of all nations participating in 1934. Courtesy Oil and Gas Journal.

the exposition. The march was followed immediately by a complicated drill routine performed by the Bugle and Drum Corps of the Joe Carson Post of the American Legion. The final act in these rites was a procession of girls wearing costumes from participating foreign nations who marched from the Red Cross Building to the speaker's stand carrying an array of flags representing the nations. They were accompanied by the A & M band playing "America."[15]

Official attendance at the opening day of the IPE was given by Manager Way as 10,852.[16] This number, reported Way, considerably exceeded the opening-day attendance at the 1930 show, which was 8,000. It appeared that Way's prediction of a complete sell-out would materialize. Shortly before the exposition opened, he had reported that four of the five exhibit buildings were completely

N.V.V. Franchot, keynote speaker in 1934. Courtesy Oil and Gas Journal.

Foreign delegates posing in 1934. Courtesy Oil and Gas Journal.

sold out and that 406 separate contracts had been signed for exhibit space. The only building with available space was the California Building, and as Way predicted it was sold out by show time.[17]

As in the past, foreign interest in the IPE was strong. A host of delegates and guests from foreign countries were seated on the speaker's stand during the opening-day ceremonies. The countries represented were Belgium, France, England, Italy, Czechoslovakia, Rumania, Portugal, Poland, and French Morocco. Delegates from these countries and their guests formed a half-circle on the reviewing stand and stood at reverent attention as President Skelly dedicated the IPE's mammoth American flag to the Joe Carson Post of the American Legion.[18]

The exhibits and displays were opened and staffed to welcome the throngs that would move through them in the next few days. These exhibits mirrored trends and developments within the industry, the most recent major trend being an intensifying of emphasis on refining and marketing. The 1934 show was characterized "by a continued movement toward greater marketing displays and by development of pumping equipment for deep wells in Oklahoma and Texas where the natural gas pressure was playing out."[19]

Space in the Refiners and Marketers Building

Russian delegates at the 1934 IPE. They remained aloof from other participants, touring the exhibits in a group. Courtesy Oil and Gas Journal.

had been sold out for months in response to the growing importance of this segment of the industry and to the emphasis being given it at the IPE. The *Manufacturer's Record* noted, "One of the most interesting developments in the oil industry has been the marketing division, and the importance of it has been recognized by the exposition directors. An entire building is given over to the marketing end of the industry and will have in it exhibits from the majority of the larger marketing companies."[20] Marketing had begun as a sideline to the grocery business. The old box and pump which had been at the rear of the store where kerosene was sold was brought to the front of the store and gasoline was added. Pumps were also set up by the owners of repair garages, and a whole industry developed and flourished.[21]

Marketing was largely neglected in the earliest expositions, which were "mostly a producer's affair."[22] Refining equipment soon began appearing, but "not until 1930 did marketing claim much of a part."[23] The Refiners and Marketers Building was erected that year, and its exhibit space was quickly sold out. The completion of this building gave the IPE its complete plant, one valued at more than half a million dollars.

The refinery division had also become increasingly important with each succeeding show, and in 1934 its main exhibit reflected the history of its development. Part of this exhibit consisted of the first refinery, built at Titusville, Pennsylvania, shortly after Colonel Drake completed his well. The counterbalancing part of the exhibit consisted of a model of a modern refinery which "uses practically every particle of the crude, even, in some instances, the smoke that goes up through the stacks."[24] Also displayed was a selection of the many by-products developed in a modern refinery.

The production division, as it had since 1923, presented its replica of Colonel Drake's well and again placed on display the rudimentary tools used in completing the well to its 69½-foot depth. The production division's contrasting display was provided by actual modern rigs with $125,000 derricks and rotary equipment which could reach depths of two miles or more. Almost every imaginable kind of derrick and drilling tool being

Marathon Oil's booth in the Refiner's and Marketer's Building in 1934. Space in this building was sold out months in advance of the show. Courtesy Oil and Gas Journal.

used in 1934 was on display at the show, and many of these were being used for actual drilling on the grounds.[25]

The transportation/pipeline division had its own spectacular display to add to the wonders on view at the eighth IPE. The Society of Mechanical Engineers had installed on the grounds an automatic pump station to show how oil could be mechanically pumped from wells. The station was valued at $90,000 and was one of the marvels of its day. It was equipped with sensors which would start the pump automatically when the oil level in the flow tank fell to a certain point and stop the pumps when the tank filled or when a breakdown occurred anywhere in the pipeline. Other exhibits in this division included the latest types of oil well pumping machinery, power units, and accessory equipment.[26]

Other features to be found among the exhibits included mammoth engines used in the oil field for pumping or drilling and pipeline pump drives which reduced or eliminated fire hazards because of their improved starting and cooling systems. These engines contained a solid injection spark ignition and operated on fuel oil, thus eliminating the need for gasoline fuel. Such engines were especially useful for wildcatting.[27]

A multitude of other exhibits were to be found as the visitor moved from booth to booth. There was wire cloth made of new corrosive-resisting alloys; blankets for the very latest clay towers made of a single piece; improved methods for applying filter cloth to filter leaves; clay towers which, owing to high heat and corrosion, could not be operated without the use of highly resistant metal screens; new drill pipe with patented joint features having 30 percent more bearing surface, eliminating rocking in threads, washouts, breaking off in threads and providing sealed shoulders; wild well capping devices used to shut in wild wells at Oklahoma City and the gulf coast; improvements in hook-ups for Christmas trees on producing wells; bricks that increased the life of furnace linings; seamless drill pipe; casing tubings; air cleaners and oil filters which lessened operating costs; faster and safer winches and pumps on special oil field trucks; new types of drilling tools which permitted easier operation and gave more accurate and better results; chain drives which reduced heavy pumping costs; plunger lifts which made flowing wells out of pumpers; sucker rods for corrosive conditions which made possible the pumping of highly corrosive wells; improved pressure regulators for all purposes; and a host of other items which were coming on-line to make the oilman's job easier, more efficient, and safer.[28] Exhibitors not only displayed these items but also gave many actual demonstrations so that viewers could visualize the equipment at work in the oil field.

W. A. Schleuter, chairman of the scientific and technical committee, had coordinated the displays of the "magical" segment of the oil industry in the Scientific and Technical Building. The star of the show in this department was the "electric eye." Schleuter said of this development:

> What this eye willl mean to the petroleum industry has never been dreamed of. It can be safely said that the most mechanical or routine operations dependent upon man's sight or touch can be done by the photoelectric cell with perfect regularity, flawless judgment and absolute freedom from fatigue. This means a tremendous saving in cost, waste and loss of time. The fact that the cell responds to light so varying in intensity and frequency and can be made to vary in linear, logarithmic and diversified curves makes it possible to adapt it to practically all branches of the petroleum industry.[29]

Schleuter was assisted on his committee by Dallas Gregory, consulting petroleum engineer of Tulsa and R.C. Beckstrom, former dean of the petroleum engineering department of the University of Tulsa. The committee doubled the space for scientific and technical exhibits over that available in 1930, with an eye toward having "the most extensive display of scientific apparatus connected with the oil industry ever gathered at one time."[30]

Less arcane demonstrations and displays were given outside on the grounds. Chief among these was the annual first aid competition which pitted first aid teams from the major oil companies against each other in tests of speed and medical knowledge. May 18 was designated as Safety Day, and it brought together teams from all over the Mid-Continent territory. The first aid event was first held at the 1925 exposition and quickly established itself as an outstanding event of each succeeding show. "First aid," said G.O. Lockwood of the Empire Companies in Bartlesville, Okla-

Mammoth engines for pumping and drilling were popular in 1934. Cooper Bessemer was a leader in this industry. Courtesy Oil and Gas Journal.

Tulsa banks enthusiastically participated in the IPE. In 1934 the National Bank of Tulsa filled its reception room with oil memorabilia. Standing in the doorway are Walter Ferguson, vice president in charge of assembling this collection, and A.E. Bradshaw (right), president of the bank. Courtesy Oil and Gas Journal.

homa, and chairman of the executive committee, "is an integral part of the training for workers in the oil industry."[31] He added, "We approach this contest with more than eighteen thousand trained first aid men in the mid continent territory and over four hundred living examples of the effectiveness of the work, men who today would be in their graves were it not for the trained ability of their fellow workers."[32]

The importance accorded safety training and first aid in the oil industry was indicated by the three-fold sponsorship for this event. It was presented jointly by the exposition, the Petroleum Safety Council, and the department of accident prevention of the American Petroleum Institute.[33]

A related but more spectacular event was a fire fighting demonstration given by the famous Kinley Brothers, specialists in extinguishing oil well fires. The *Tulsa Daily World* referred to this event as "perhaps the most spectacular thing we've ever seen…. Tulsans who haven't seen it will be foolish

A safety-first demonstration, sponsored by the IPE and the Petroleum Safety Council, along with the department of accident prevention of the American Petroleum Institute, was a highlight of the 1934 show. The event took place on Suppliers' Road. Courtesy Leslie Brooks.

to miss the second demonstration....It's a hot sure-fire drama filled with high pressure entertainment."[34]

The chief actor in this drama was M.M. Kinley who was generally recognized in the oil industry as the greatest oil well fire fighter in the world by virtue of having extinguished a Rumanian oil well fire which had defied European experts for more than two years and had cost more than a million dollars. Its greater cost, however, was the lives of 14 Rumanian oilmen, some of whom were working on the rig when the fire broke out and some of whom came later to fight the fire. Kinley scored a spectacular success in putting out the fire quickly.[35] Before the exposition presented them to the public, only a privileged few had seen Kinley and his brother Floyd at work in their asbestos suits, armed with charges of gelatinized nitroglycerin.

An oil derrick was raised for the purposes of the demonstration, and gas under high pressure was piped to its floor. When the gas was ignited, a tower of flame engulfed the derrick and shot about a hundred feet into the air while Kinley struggled into his asbestos suit and prepared for the battle. After about an hour, during which Kinley assembled his equipment and prepared his nitroglycerin charges, the rig collapsed into a heap of glowing rubble through which the flames from the burning gas continued to spew. Working methodically, Kinley pulled the seething metal clear of the flames with grappling hooks and edged closer with his blasting charges at the ready. At precisely the right moment, he hurled the charges at the base of the flame, setting off a deafening blast which snuffed out the roaring gas fire and left the scene smoldering with wisps of smoke and drifting vapors. The tumultuous applause of the spectators almost equalled the thunderous clap of the nitroglycerin.[36]

So popular was Kinley's demonstration that spectators crowded into the viewing stands in numbers greater than they could bear. During the evening performance of May 16, a section of the overburdened grandstand collapsed, tumbling spectators in every direction but causing only minor injuries. Several claims were filed against the corporation, and final settlements amounting to $1,201.50 were paid.[37]

A different kind of entertainment was provided by a great horse show scheduled to coincide with the exposition. The *Southwestern Horseman* predicted that the show would be "one of the greatest horse shows in the southwest this spring...inasmuch as some of the outstanding stables in the country are located in or around Tulsa and Oklahoma City."[38] Exhibitors entered the three-day show from the contiguous states of Texas, Kansas, Missouri, and Colorado to vie for approximately $3,000 in cash prizes. Handsome trophies were among the prizes awarded in the five-gaited stake, the three-gaited stake, and the roadster stake. The *Horseman* urged equestrians from all over the Southwest to participate in the horse show and in the exposition.

Such entertainments as the horse show paled in comparison to an entertainment feature which sought to encompass the entire panorama of oil as a valuable commodity in the modern world. This was the "History of Oil" pageant presented in the pavilion May 14-16. Presented in five acts, the pageant incorporated music, dance, and acting in an effort to summarize the uses to which oil had been put from Biblical times up through the diamond jubilee year of the industry.[39] Miss Isabel Ronan from Tulsa Central High School, acclaimed as a teacher of speech and theater, was enlisted to direct the lavish production. Music was provided by the Central High School orchestra, the A Cappella Choir, and the St. Cecilians, directed by Albert Weatherly, George Oscar Brown, and Reed Jerome respectively. Lavish scenery and costumes were brought from Kansas City, and a three tiered stage was constructed to allow simultaneous presentation of several tableaux. Neither expense nor effort was spared in the preparations.

Act I of the pageant portrayed the uses of oil in ancient religious rites, concluding with an intricately choreographed "Dance of Fire." Act II presented "Petroleum as a Building Material" and showed how the substance was used in ancient Egypt. Act III carried the story forward with "Petroleum as Illumination," and Act IV illustrated "Uses in Early American Times." Act V brought the story up-to-date, depicting the 75-year development of the oil industry and concluding with a "Dance of the Machine Age."[40] Unfortunately, the pageant did not attract great interest, perhaps because it competed for attention with many real-life processes and artifacts which it could present only symbolically.

Many of those who viewed the pageant, particularly Act V, felt nostalgic stirrings as the actors and musicians dramatized the history of the modern oil industry, for they had lived it. These were the Old Timers who were reliving their memories at the IPE through the activities of the Pioneers Association, then in its fourth year as an official part of the exposition. Of course, the Old Timers were on hand mainly to see who from their number would be picked for the honorary titles of "Grand Old Man of the Petroleum Industry" and "Pioneer of Pioneers." The "Grand Old Man" title went to N.V.V. Franchot of Olean, New York, who began his career as a gauger in 1875 in the Pennsylvania oil fields. A. W. Gordon of Tulsa was named "Pioneer of Pioneers." He also had entered the oil fields in Pennsylvania, working as a rigger.[41]

The Old Timers, along with all other participants in the eighth International Petroleum Exposition, found the show to be "the greatest and most effective in the history of the industry, with the consensus of opinion being that it definitely marked the turning point for the nation's largest industry."[42] This upturn was the result of the natural exuberance and optimism of oil men and improved economic conditions in the nation generally. Reports of an increase in business activity were becoming common. State sales tax revenues were increasing steadily, and the oil industry was heartened when Secretary of the Interior Harold Ickes, who also served as federal oil administrator, announced in May that Oklahoma's crude oil allocation for June would be increased by 11,300 barrels daily. On the basis of dollar oil, the increase would mean a leap of $40,000 a day in the industry's income.[13] An equally strong indicator was the upswing in sales by exhibitors at the IPE where sales of up to $80,000 on a single transaction were recorded. The judgment of the *Oil and Gas Journal* was, "No phase of oil interest has been overlooked in the arrangements by the exposition management to make the 1934 International Petroleum Exposition at Tulsa the most comprehensive and interesting yet staged. This, we believe, will be the verdict of every visitor."[14]

1936

Tulsa became "the Mecca of thousands of oilmen from all parts of the world"[15] during the eight days from May 16 to 23, 1936. Even before the exposition opened on Saturday the 16th, oil men had pressed into Tulsa in great throngs to prepare for the show. By the Wednesday before the show, registration was heavy for the various professional societies which had planned their meetings to coincide with the IPE.

The *Oil and Gas Journal* reported that "advance registration indicates that Tulsa's capacity as host will be extended to its limit when all participants in the show finally arrive."[16] The *Journal* made its projections partially on the great numbers of oil men who were appearing for the pre-exposition meetings of such groups as The Natural Gasoline Association of America and the various divisions of the American Petroleum Institute. These associations were joined by the following groups in preparing for the IPE by convening their memberships: the Institution of Petroleum Technologists; the Purchasing Agents Association of Tulsa; the Oil Company Buyers Group of the National Association of Purchasing Agents; the Junior Chamber of Commerce Young Men's Association; the Independent Petroleum Association of America; the National Stripper Well Association; and the Association of Drilling Contractors.[17]

In order to foster cooperation and unity among these various groups, 18 of them jointly planned a gala banquet for Thursday evening, May 21. Charles L. McMahon was selected chairman of the Tulsa committee on arrangements for this event, and J. Wood Glass, president of the Northeastern Oklahoma Stripper Well Association, was chosen to preside at the banquet. Toastmaster for the event was Charles F. Roeser, president of the Independent Petroleum Association of America. The banquet speaker was John M. Lovejoy, president of the American Institute of Mining and Metallurgical Engineers who talked on "The Relation of the Technical Problems of the Petroleum Industry to its Economic Problems."[18]

This banquet and related activities were reminiscent of the old-time Congress, and much satisfaction was expressed over the sense of community that grew out of the event. Oil men once again were feeling the need to come together outside of their specialities and share common interests and problems. The divisiveness that had crept into previous expositions over the issues of overproduction and proration no longer was in evidence; cooperation and communication were the keynotes in 1936.

No contingent within the IPE was more gratified by the exuberant sense of cooperation than its officers and directors, their number having grown to 80 by 1936. Continuing as officers of the corporation were W.G. Skelly, president; Frank J. Hinderliter, vice president; C.H. Pape, treasurer, replacing A.W. Leonard who resigned after the 1934 show; William B. Way, general manager; and Reid S. McBeth, secretary (William Holden moved to Fort Worth, Texas, and relinquished his office). McBeth himself would decline a second term in 1938 and be replaced by H.R. Powers, manager of the oil department of the Tulsa Chamber of Commerce.[19] Members of this group met almost daily, either as a committee of the whole or in hurried gatherings of two or three during the final weeks of preparation for the great show.

Nor was there any lack of activity on the exposition grounds. Crews from hundreds of manufacturing and supply companies were scurrying over the 23-acre site to complete the installation of exhibits which would cover every phase of the industry "from preliminary exploration of wildcat areas to the finished product as it goes to the consumer."[50] The variety and scope of the exhibits exceeded that of any previous show. When all were finally in place and ready for viewing by the thousands of visitors, more than $1 million worth of oil paraphernalia had been assembled.

A conspicuous emphasis in the ninth show was given to scientific and technical equipment, and this segment of the oil industry was represented

Oklahoma Governor E.W. Marland welcomes guests to the 1936 IPE. Courtesy Oil and Gas Journal.

by thousands of technical employees from research departments, plants, and oil fields as well as by students from university departments of engineering and research. The latest and most sophisticated instruments and processes were demonstrated, and everyone who made his living—or hoped to—searching for the elusive hydrocarbons wanted to be on hand personally or by proxy. The sense of anticipation ran high when, on May 16 at 9:00 a.m., the assembled dignitaries on the exposition speakers platform opened the ninth International Petroleum Exposition and Congress.[51]

Principal speaker for this ceremony was W.A. Irvin, president of the United States Steel Corporation, whose presence symbolized the continuing and growing interdependence of America's industrial giants, steel and oil.[52] Greetings were extended to participants and visitors by Tulsa Mayor T.A. Penny, Oklahoma Governor E.W. Marland (who had founded Marland Oil Company), and IPE President Skelly. Following the speeches and remarks came a display of pageantry to conclude the opening ceremonies. Aerial bombs darted skyward and thundered over the crowd, while 20 cowboys riding four abreast paraded before the speaker's stand to the martial strains of the Oklahoma Military Academy band. The conclusion of the ceremony was provided by Indians doing a ritual stomp dance.

As excitement at these performances subsided, the great crowd began to form queues along the lanes and walks running through the many exhibits. Beginning with the second exposition in 1924, the names of famous persons, cities, or oil fields had been given to the passageways, whether walkways or roads, through the exhibits. Some of the names were Drake Drive, Diesel Drive, Rocky Mountain Road, California Avenue, Seminole Street, Skelly Drive, Cushing Lane, Old Timers Avenue, Gasoline Alley, Oklahoma Drive, Silver Lane, Signal Hill Street, and Avenida Maracaibo.[53] For the ninth time these passageways were filled with awed spectators admiring the wonders of the oil industry, and the wonders were larger, more animated, better illustrated, and more beautiful than ever before.[54]

There was nearly a mile of displays in the 476 booths housed in the IPE's five giant exhibit buildings where all facilities had recently undergone renovation and beautification. New floors were installed in several places and electrical, water, and gas facilities were provided for all exhibit booths. Inside many of the booths the handiwork of professional designers was in evidence, and in each building the rough corrugated walls had been lined with sateen to improve the overall esthetic effect of the exposition. All of the buildings had been repainted inside and out before the opening of the show. Many exhibitors had also ordered chromium-plated furniture for their booths, as well as other amenities that enhanced the comfort, convenience, and appearance of their exhibits. Way commented, "This is the first time we have had a call for such fine furniture."[55] All of the show's participants wished to contribute to the magnificence of the exposition.

The exhibits themselves were bigger, better, and more numerous. More standard drilling rigs of the rotary and cable varieties were on the grounds. One rig was a "combination rotary driven by diesel power, steam, and electricity."[56] Also on display were gas regulator valves, arc welding machines, oxygen machines, seismographs, steam drilling machines, fire fighting machines, underground surveying instruments, pressure gauges, explosion-proof motors, air filters, magnetic telephones, and other pieces of equipment which contributed to the search for, production, and processing of oil.[57] As Way declared, "There has been more showmanship put into cold machinery than ever before. The beauty of the exhibits and the way in which machinery has been glorified has been impressive."[58] Way commented further that he believed the World's Fair at Chicago had influenced the improvement in exhibits, "making them more beautiful and actually putting in working exhibits....This means that the exposition itself is having to dress up the plant a bit."[59] The beautification efforts, said Way, would probably take between five and 10 years.

The glorification of machinery was counterbalanced by human interest activities designed to emphasize the importance of the human element in the oil business. Originally the main focus of this aspect of the show was on the pioneers, but the exposition had also begun to recognize its future pioneers, its young leaders who would determine the destiny not only of the oil industry but also of the IPE. On May 18, 1936, the Tulsa Junior Chamber of Commerce held its "Young Oil Executives Program" to orient these young leaders in the industry they would lead. Registration took place from 10:00 a.m. to noon in the third-floor lobby of the Tulsa Building. On Tuesday the 19th, the group met for a luncheon in the Chamber of Commerce dining room on the fourth floor of the Tulsa Building at Fourth Street and Cincinnati Avenue. Joe Gill, chairman of the "Young Executives," presided at the meeting, and Fred Holmes, president of the Tulsa Junior Chamber of Commerce welcomed the visitors to Tulsa and the IPE. Harry Krusz, former secretary of the United States Junior Chamber of Commerce, introduced guests at the tables and turned the program over to Kirby Cowan from the Dallas Junior Chamber of Commerce who brought "Words from Texas." The featured speaker for the luncheon was Charles F. Roeser, president of the International Petroleum Producers' Association. Concluding the formal presentations at the luncheon was Leslie Brooks, chairman of the IPE publicity committee, who announced the activities for the rest of the event.[60]

One of these activities was a field trip to Beggs, Oklahoma, for a seismograph demonstration presented by the Seismograph Service Corporation. That evening the young men were entertained at a dance held in the Crystal Ballroom on the 16th floor of the Mayo Hotel. Also available without charge to young executives who presented their registration identification was the Avery Gold golf course, which proved to be highly popular with the young oil men—who were "notoriously fond" of golf according to William Sherry.[61]

At the other end of the spectrum among the human interest activities of the IPE was the Old Timers program. The general Old Timers committee was headed in 1936 by Victor F. Barnett, who was assisted by members L.E. Smith, Dal Dalrymple, Paul Hedrick, A.W. Leonard, N.V.V. Franchot, H.R. Gruber, Floyd Swindell, James McIntyre, Barney Horrigan, and W.G. Skelly. Something new in 1936 was the projected installation on the exposition grounds of a memorial to the Old Timers. It was to consist of an engraved plaque and weather-sealed registration book of Old Timer membership. This also was the year when the Old Timers appropriated for their headquarters the replica of the Drake Well which had been moved to the center of the exposition grounds and "fitted up as a clubhouse for these men."[62]

Two medals were awarded in 1936, one to the "Grand Old Man of the Industry" and one to the "Grand Old Man of Tulsa." The criteria for the selections were mainly seniority but not entirely so. The "Grand Old Man of the Industry" was required to have at least 50 years in the industry, and the historical subcommittee of the Old Timers committee decided to add the statement that he must also have contributed to the industry by personal example, inspiration, invention, or discovery. The same requirements applied to the title "Grand Old Man of Tulsa" except that the in-service time was reduced to 30 years, 20 of which had to be spent in Tulsa.

When the committee concluded its lengthy process of collecting credentials, reviewing, discussing, and voting, two Pennsylvania natives emerged with the coveted titles. Martin H. Mosier, who resided in Los Angeles at this time, was named "Grand Old Man of the Petroleum Industry," and W.J. Stewart was named "Grand Old Man of Tulsa." The "Pioneer of Pioneers" title went to William H. Peiffer of Carthage, Missouri. Each man received a gold engraved medal showing his title, and other Old Timers received gold lapel pins indicating their distinguished standing in the oil industry.[63]

Martin Mosier, the "Grand Old Man of the Industry," claimed 60 years service, having begun his career as a well pumper in 1876 at Petrolia, Butler County, Pennsylvania, when he was 20 years old. He followed the oil business throughout the oil states, working from Bolivar, New York, to the Pacific coast. He made a special study of natural gas and participated in building the first natural gas pumping station in 1881. Later he became an oil operator in Tulsa and drilled his first well at Glenn Pool. Still later, he moved his operations to California, but at the time of his election he continued to oversee his interests in Oklahoma.

W.J. Stewart, Tulsa's "Grand Old Man," also began his active service in the oil industry in Butler County, Pennsylvania, where he helped his stepfather as a pumper and in a variety of other oil field duties. In 1876 he moved to McKean County, Pennsylvania, and worked "on his own" as a roustabout. From these beginnings he continued to work in the oil fields until he became an independent producer, entering the Sisterville, West Virginia, Field and helping to organize the Melrose Oil Company. In 1907, the year of Oklahoma's statehood, he came to Tulsa and joined the Devonian Oil Company with which he was still active at the time of his election.[64]

The "Pioneer of Pioneers," William H. Peiffer, 93, a Civil War veteran, took his first job in the oil industry as a rig builder and driller on Stephenson Hill near Titusville, Pennsylvania, the birthplace of the oil industry. Unlike other honorees, Peiffer did not remain in the oil business, moving with his family to Missouri in 1882 and leaving the oil fields behind.[65] When these gentlemen received their awards, they joined a select group of 14 pioneers who had been singled out for special notice since the first awards had been given at the 1927 exposition.

One IPE feature which did not increase in 1936 was foreign participation. Before the depression years as many as 20 foreign nations had sent some 80 delegates, but this number declined in 1934, and in 1936 only 14 foreign countries were represented. The depression had far reaching effects, and many nations were forced to cut back on all but essential expenses. Nevertheless the international delegates committee pressed forward with its plans to entertain its guests "in style." The 1936 committee was headed by John Silsbee, who was assisted by members C.E. Buchner, Edward C. Lawson, George F. Martin, Paul M. Raigorodsky, and Earl Sneed. On May 19 the committee entertained all of the foreign delegates at a reception and banquet at the Tulsa Club. The reception began at 6:00 p.m. and the banquet an hour later. Nelson K. Moody served as toastmaster for the event, and the main address was given by Harold B. Fell.[66] Tours, parties, and more receptions followed on succeeding days of the exposition to round out a busy and involved schedule of hospitality and orientation for the foreign visitors.

Entertainment and hospitality were available to all participants in the 1936 show, but as had been determined by the directors several years previously, these were kept secondary to the commercial and professional purposes of the exposition. A Frick-Reid advertisement in the May 14, 1936, issue of the *Oil and Gas Journal* announced:

> The old side-show barker and the medicine Fakir played to a credulous and willing audience. Their stock in trade was blatant bally-hoo. Contrast that with the scientific intent of the International Petroleum Exposition at Tulsa May 16-23....Not that the Tulsa show won't have its festive atmosphere—a time and place for renewing old friendships and making new contacts.[67]

A favorite place for oil men to renew old friendships and make new contacts was on the golf links, and entries in the 1936 IPE Golf Tournament reached a new high. Frank Gray, golf tournament chairman, announced "This should be the best tournament in Oklahoma history if the men we expect are here. I believe it will take a score of 77 or better to qualify in the championship flight."[68] Former Trans-Mississippi champion Jimmie Manion of St. Louis, Missouri, was on the entry list "with a flock of other titlists and near-titlists from a half-dozen states."[69] Spectator entertainment was provided by Major Gordon W. "Pawnee Bill" Lillie and his traveling rodeo. His 125 cowboys and cowgirls put on a thrilling display of bronco busting, trick riding, and steer wrestling in the IPE pavilion before capacity crowds.[70]

The pavilion was also the site for the second IPE appearance of the fire-fighting Kinley brothers whose 1934 demonstration of techniques for extinguishing oil well fires was a high point of the eighth exposition. For their 1936 demonstration Myron and Floyd Kinley asked that an 84-foot derrick be erected and equipped with the usual supplies and tools for drilling. Gas supplied by Pure Oil Company was again used for the simulated explosion and fire. After the first explosion was touched off, the Kinleys went deliberately about the business of assembling their equipment, pulling on their bulky asbestos suits, and waiting for the derrick to cave in upon itself amid billows of sparks and sheets of flame, a sight which brought a gasp from the thousands of spectators.[71]

When the derrick and tools lay in a white-hot pile of rubble, Myron Kinley threw his grappling hooks into the sizzling mass and began dragging away disfigured hunks and shards of the ruined derrick to prevent reignition of the still-escaping gas. His final operation was to lay a nitroglycerin charge within a few feet of the inferno, move away, and set off the charge, extinguishing the flame by choking off its oxygen supply for an instant.[72] The crowd applauded the Kinleys thunderously after sitting for a moment in stunned silence.

Fire and smoke of another kind hung over the IPE in 1936, for the show, by virtue of its unparalleled worldwide success, had attracted competition, the Oil World Exposition held in Houston, Texas. That show had begun in 1930 on the West Coast as the Oil Equipment and Engineering Exposition. Three shows were held there—in 1930, 1931, and 1932—and participation was respectable, at least in comparison to the first few presentations of the IPE. By the 1930s the IPE had advanced so far from its beginnings that the new show did not compare favorably in any way with it. The West Coast show moved to Houston in 1933 and was renamed the Oil World Exposition. At its new site it made six appearances, 1933-1937 and 1939. Its managers attempted unsuccessfully in 1936 to persuade officers of the IPE to assume management of the Houston effort. The idea was that the two shows would alternate, thereby complementing each other.[73] Tulsa's oil leaders rejected the proposal out of hand, knowing that the oil companies would not be disposed to shuttle between Houston and Tulsa on alternate years.[74]

The Oil World Exposition people then took another tack. They began attending the IPE in some force and attempting to lure exhibitors to Houston, but their efforts were largely unsuccessful. They continued to draw exhibitors from Houston and the Gulf Coast generally, but their proselytizing efforts paid slim rewards. The failure of these efforts led OWE managers to decide to engage in a showdown with the IPE in 1942, and they announced the opening of the Houston show for early April, mere weeks before the IPE was to open in May.[75] What neither group of managers could have known was that on December 7, 1941, Japanese bombers at Pearl Harbor would cancel both shows. The IPE directors met on December 8 in emergency session and announced that the 1942 oil show in Tulsa would be cancelled so that the oil industry could throw its full resources into the war effort. The Houston show also was cancelled. During the interim between the tenth and the eleventh expositions, the Houston show faded into oblivion. The IPE in 1936, however, appeared to store up momentum which was pressing forward powerfully in response to the revival of profits within the oil industry.

1938

The IPE increased its size by 50 percent in 1938 with the addition of $250,000 worth of new facilities. Two new exhibit buildings were added along with a new cafeteria, a new Hall of Science, a new office building, and seven new buildings constructed by individual exhibitors.[76] Among these were the Oilwell Supply Company, the International Harvester Company, W.C. Norris Company, Gaso Pump and Burner Company, Petroleum Engineer Publishing Company, Bovaird Supply Company, and Hinderliter Tool Company. Firms which already had permanent structures on the grounds were National Supply Company, Bridgeport Machine Tool and Supply Company, Continental Supply Company, Gates Hardware Company, and Frick-Reid Supply Company. The new private structures added 27,552 square feet of exhibit space under contract to private companies. Additional ground space also was added when the exposition management decided to move the fence bordering the east side of the grounds to the property line.[77]

As this new exhibit space was opened, load after load of supplies and equipment began pouring into the exposition grounds. Two days before the show's opening, 236 such loads had been deposited at various exhibit sites, and company officials were busily attending to last-minute arrangements before their exhibits were unveiled.[78] The grandiose preparations for the tenth International Petroleum Exposition moved many participants and observers to reflect on growth and progress in both the exposition and the oil industry since King Petroleo first decreed the existence of his realm in 1923. Axtel J. Byles, president of the American Petroleum Institute, prepared an article for the *Tulsa Tribune* in which he reviewed the 15 year interim between the first IPE and the tenth:

> Many thousands of oil men who will visit the nation's greatest oil show can remember the first of these events which took place nearly fifteen years ago. They came to learn and to inspect the many exhibits of the most modern equipment then available to the petroleum industry and to take advantage of the improvements offered.
>
> Today these oil men are back in Tulsa for the same purpose. They are still learning despite the enormous fund of knowledge that has been added to the aggregate of petroleum technique and science during the fifteen years that have lapsed since that first International Petroleum Exposition of 1923.
>
> A comparison between modern equipment and methods and those of fifteen years ago would yield approximately the same sort of contrast as comparison between an up-to-date streamlined car and the "tin lizzie" of 1923. Much of the equipment that crowds the many acres of exhibit space today was not even a vision in the minds of its developers back in 1923.
>
> Even fifteen years ago the petroleum industry was a great industrial enterprise with an invested capital of $8 billion. Annual production of crude oil amounted to nearly 750 million barrels and to serve the 15 million motor vehicles then in use it produced nearly 200 million barrels of gasoline.
>
> Since 1923 the industry's motorist-customers have doubled in number, and users of other petroleum products proportionately have increased in their requirements. The annual output of crude has been increased by more than 500 million barrels, much of it coming from the great new fields discovered by vastly improved methods of search. Some 300,000 wells have been drilled and of this number nearly 200,000 have been productive of oil. The deepest well producing in 1923 did not exceed 7500 feet. Today the deepest producer brought in recently in California penetrates the earth's crust twice that distance.[79]

The *Tulsa World* also presented an encomium to the exposition and the oil industry:

> There is not in the active United States today a more genuinely and effectively busy place than the International Petroleum Exposition tract. Four days and four nights remain for making ready the world's greatest single-industry exposition.
>
> Mighty changes and great extensions have come to the oil industry, in all its branches, since the 1936 exposition. The machinery is finer, more powerful; the uses for all the 1,600 derivatives of petroleum have been found; the business is going three miles down for oil and controlling the operation from the surface has been advanced. The educational feature of the exposition, well outside the range of routine and commercial exhibits, can be described only in superlatives.
>
> The large tract is now almost covered with permanent buildings and exhibit bases. There will be upwards of five hundred exhibitors, each with notable progress to demonstrate to all comers. The fact that twenty nations are contributories is eloquent of the scope and drawing power of the exposition.
>
> When considered, the exposition is a grand opportunity, a brilliant expression of enterprise, a fine school, a delight to the senses, and a special time for exulting in the power and prowess of the oil industry. All of us should attend the exposition—many times.[80]

Tulsa was truly "exulting in the power and prowess" of its great industry. Its preparations for the tenth IPE could only be described as exuberant. All streets in the downtown business section were decorated with overhead banners, and downtown office buildings were kept ablaze with light and special electrical decorations. Plans were made for welcoming out-of-town visitors with bands and marches through the downtown streets. Mayor Frank Martin of Oklahoma City was extended an invitation to assume "host" duties for Tulsa and the exposition. He accepted in the ironically playful spirit in which the offer was made, saying, "I'm a strong booster for the exposition. Everything that helps the state's rural communities helps Oklahoma City, so I'll be on hand."[81]

Cooperation came from every quarter. Eighty Oklahoma cities and towns provided honorary

Flag-raising ceremony on opening day of the 1938 IPE. Courtesy Oil and Gas Journal.

H.C. Merritt, vice president of Allis-Chalmers, speaking at the opening of the 1938 IPE. Courtesy Oil and Gas Journal.

"hosts" for the thousands of visitors thronging into Tulsa, and each of these dignitaries was provided with a "host badge" to signify his or her official capacity. Cities on all sides of the Oil Capitol sent their school bands to provide music for the occasion. The entire state seemed caught up in the gala spirit that marked the preparatory stages of the tenth IPE. The incoming crowds strained to the limit the community facilities of Tulsa, particularly the municipal airport which recorded its heaviest flow of traffic in seven years. One newspaper report stated that, "Private ships and regular schedules of Hanford and American Airlines poured visitors into Tulsa for the International Petroleum Exposition."[82] The nationwide communications facilities of the National Broadcasting Company were brought into play for a network-wide broadcast of IPE activities on the first Saturday of the show. The program was titled, "The March of Petroleum."[83]

The show's opening on Saturday afternoon "was auspicious."[84] A record-breaking crowd poured through the turnstiles at the main gate, and the skies which had threatened rain at daybreak were clear and bright. Drake Drive was crowded with participants for the official opening ceremonies: cowboys and cowgirls from the exposition's rodeo; a troup of Boy Scouts carrying the colors of all nations; and throngs of spectators.[85] Gasps were heard as the IPE's giant American flag was unfurled amid a crash of music and exploding aerial bombs. Even though several foreign delegates were introduced, the spirit and activities of the opening ceremonies could be described as nothing less than a pure, proud celebration of Americanism, and this feeling was heightened by the opening speeches.

President Skelly spoke with such nationalistic pride that he sounded almost like a Fourth of July speaker. "This is a serious show for serious people," he said, "and yet these grounds are filled with the romance of American industry. On these grounds are a thousand believe-it-or nots."[86] He went on to say, "This industry was developed in America; it is typically American today. If people of the United States could see this tremendous array of equipment, they would say that the United States is safe and we are going forward faster in the future."[87]

The keynote speaker for the opening ceremonies was A.E. Barit, president of the Hudson Motor Company of Detroit, who had been brought to Tulsa by William Duvall of the Duvall Motor Company. Barit also sounded a strong note of optimism and pride in the resourcefulness and strength of America's oil industry: "I am still gasping with amazement at this exposition. It is great not only because of its size and its scope; it leaves me with the feeling that the development of a basic natural resource is in competent hands."[88]

Speaking with even greater force, H.C. Meritt, vice president of the Milwaukee-based Allis-Chalmers Company, eulogized the IPE, Tulsa, and Oklahoma as prime examples of the American system at work: "If it were possible for the people who believe the United States is in a tailspin to come to Oklahoma and to Tulsa and to see this exposition, they would regain their faith in our institutions and our form of society."[89]

The sentiments expressed by these speakers

were heartily applauded by the array of dignitaries accompanying them on the speaker's platform. Among these notables were J.A. LaFortune, vice president of Warren Petroleum Company and "host for the day"; Oklahoma Governor E.W. Marland; Robert S. Kerr, then president of the Kansas-Oklahoma division of the Mid Continent Oil and Gas Association; H.R. Gruber, chairman of the Old Timers committee; John Champion, a popular early-day oil man; Clarel B. Mapes, secretary of the Mid Continent Oil and Gas Association; E.H. Moore, oil producer; Jack Powers of Bethlehem Supply Corporation; John Rogers, president of the Tulsa Chamber of Commerce; Frank Hinderliter, manufacturer; Alf G. Heggem, manufacturer; Almond Blow, vice president of Amerada Petroleum Corporation; all directors of the IPE; C.H. Pape, Selby Oil and Gas Company; William E. Yunker, Allis-Chalmers Company; and E.L. Kirkpatrick, exhibitor.[90]

The opening ceremonies came to a close under billows of smoke produced by exploding aerial bombs and the scream of airplane engines as a dozen planes from the Spartan Dawn Patrol dipped in salute to the largest American flag ever flown in the United States and to the International Petroleum Exposition.[91] The "World's Fair of the Oil Industry" was on again. The high excitement generated during the preparatory activities and opening ceremonies of the tenth IPE carried over into the rest of the show, especially into the many social activities. Movie stars figured in these events when a group of IPE officials gathered at Tulsa Municipal Airport to greet Richard Arlen, Rochelle Hudson, and Robert Cummings during a 17-minute stopover in Tulsa. The newspapers made much of this brief conference, proclaiming that "Movie Stars Lend Selves to Oil Show Spirit."[92]

Entertainment was again provided for young men who hoped to make their careers in the oil industry. As guests of the Tulsa Jaycees, the young men took tours to inspect seismograph operations and to see a refinery in full operation before being treated to a stag party followed by a dance in the Topaz Room of the Hotel Tulsa.[93] Another group singled out for special hospitality was the out-of-town press corps. The reporters were given a splendid stag buffet dinner on Saturday evening on the roof garden of the Tulsa Club. Their hosts were W.G. Skelly, W.B. Way, and Leslie Brooks, with Glen Condon serving as master of ceremonies. The following Wednesday the reporters were the guests of P.C. Lauinger at a luncheon at Southern Hills Country Club, followed by rounds of golf for all those who wished to play.[94]

Hundreds of IPE guests were entertained by the drilling contractors of Tulsa and surrounding towns at a party staged in the Coliseum near the exposition grounds. No special invitation was needed for the many guests who had attended the party during the 1936 IPE, nor for the friends they brought with them in 1938.[95] The party organized to honor the IPE's international participants became a mammoth affair in response to the contractors' general invitation.

On the exposition grounds, musical entertainment continued nonstop. Twenty full bands totaling 1800 musicians alternated with each other to provide entertainment at all times during the show. Area high schools and colleges eagerly complied with band committee chairman E.C. Vickers' request that they send their musicians to the exposition. Tulsa Central High School also sent its orchestra, under the direction of Albert Weatherly, to play a concert near the Cafe de Petrol on Tuesday evening. The Oklahoma Military Academy, the University of Tulsa, and the Veterans of Foreign Wars also sent groups of musicians.

Professional entertainers performed inside the Cafe de Petrol. Joan Brooks, a singer who had made her debut in Tulsa with the Olson and Johnson vaudeville comedy act and had then gone to New York, was appearing with "Whispering" Jack Smith. Both entertainers were enroute to Hollywood when the exposition opened, but they postponed their engagements to appear at the Cafe de Petrol. The cafe also featured the Larry Lee orchestra for dining and dancing entertainment.

An event which proved to be even more popular than the entertainment features but which was strictly serious in intent was the big demonstration by the safety teams on Drake Drive, held on Friday, May 20. Some 30 teams representing the major oil centers of the nation participated in the all day affair.[96] The meet followed an all-day safety conference held on May 19, during which numerous addresses were made on accident prevention and safety problems. The two-day program was sponsored by the Mid Continent

Petroleum Safety Council which, long before the meet, commissioned the United States Bureau of Mines to train 18 judges to preside over the various events in the meet. Winning teams were presented with silver and bronze medals and with first aid equipment provided by safety equipment manufacturers and other companies. The coveted awards were taken by the first-place team from the Texas Pipeline Company of Okmulgee, Oklahoma, and the second- and third-place teams from the Texas Company Refinery of Tulsa. The awards ceremony took place in the dining room of the Tulsa Chamber of Commerce.[97]

The major attraction of the IPE in 1938 was the equipment, supplies, and tools used to find and produce oil. Dazzled visitors to the exhibit grounds moved through a fantasy of more than a thousand displays of the latest marvels of the petroleum industry. These "tools of the trade" were valued at more than $10 million.[98] Among the displays were 402 exhibits featuring "drilling and production, advances in drilling, exploration, transportation, high pressure boilers, and super heaters and vertical engines with separate drives for rotary tables."[99] There also were demonstrations of the "economical use of steam, circumferential field welding of casing, water flooding, repressurizing, a new stage method of acidizing, wlidcat exploration, and "slim hole" drilling to 4000 feet.[100]

The Franks Manufacturing Company of Tulsa developed and introduced a truck-mounted drilling rig, the first of its kind, capable of making a 6.25-inch hole to 6000 feet. Prior to the exposition, the unit had been used to drill a 4380-foot well in West Texas, then was moved in two days to a site 700 miles away in Kansas. The portability of this rig, with its four-hour set-up time, was a significant advance that could save some $10,000 per well average.[101] The Franks Company also distinguished itself in 1938 by striking oil on the exposition grounds on the seventh day of the show. The well, which came in at 540 feet, captured the interest of reporters, and within a short time stories about it were going out over the news wire services and radio networks. Magazines of national circulation picked up the story, and several newsreel firms sent cameras and crews to record the discovery.[102]

Other exhibits in the drilling and production division included heavier pumping equipment for deep wells, a packer requiring no anchor, subsurface pressure studies to gauge the life of wells, and a portable core laboratory service. The emphasis seemed to be on equipment allowing deeper drilling, and much was made of a unit capable of boring up to seven miles into the earth. This surpassed "anything even dreamed of" previously and was "more than twice as far down as the deepest well in the world, 15,004 feet at Wasco, California."[103]

There were 317 different companies exhibiting in the transportation division. They had a wide array of advances in facilities and equipment for handling pipe, trenching machines, methods for preventing corrosion of pipe, and cleaning and coating pipe, as well as cutting the cost of its upkeep. Protection of pipelines was shown along with a process of welding which cut construction time and cost. Other displays featured valves, control equipment, engines, booster plant equipment, and safety and communication devices.[104]

The refining division attracted 303 companies to exhibit the latest developments in the processing of oil. Emphasis was on greater efficiency, higher recovery, economy, and safety. The science of polymerization was demonstrated along with the cracking and reforming of butane and pentane molecules, vapor saving devices, vapor plant recovery development, processes for higher octane gasoline, desulphating of natural gases to reduce hydrogen sulphide content or complete elimination of sulphides. The processing of lubricants also was shown along with machinery for the latest refining processes.[105]

The methods of delivering petroleum products to customers were shown by 208 marketing companies. On display were the latest machinery and equipment which might interest independent jobbers, bulk station owners, and operators of service stations. Conveniences designed for the benefit of customers also were shown, including the latest type pit and pump island lighting units and service equipment such as pressure grease guns, loading racks, dispensing units, unloading, transfer and pump motors, all types of gasoline hose, portable transmission and differential lub oil dispensers, new and improved equipment for bulk stations, and new devices for improving the function and convenience of free service hoses.[106]

Both the quantity and quality of exhibits were

W.G. Skelly and Hal Gruber congratulate A.W. Gordon for his 60 years of service in the oil industry. At age 96, Gordon was the oldest participant in the 1938 show. Courtesy Oil and Gas Journal.

awe-inspiring, signifying an industry with intense professionalism and vigor, an impression perhaps empasized most strongly in the exhibits in the Hall of Science. The display in 1938 was "the largest...ever seen at any oil show."[107] Dr. Gustav Egloff of Universal Oil Products Company of Chicago headed the planning committee for the Hall of Science, and his group assembled exhibits and working models costing more than $250,000, representing the "foremost museum of science and industry devoted exclusively to petroleum and its products."[108] Forty companies, institutions, agencies, and bureaus, including the Chicago Museum of Science and Industry, contributed items and displays. Three times more space were required than in any previous year to house displays showing prospecting, drilling, transportation, refining, marketing, and industry economics.[109]

Machinery, equipment, supplies, and techniques were popular in the IPE in 1938, but the human element held its own appeal. Of the Old Timers contest, the *Tulsa World* reported, "White headed men with horny hands and halting steps were moving slowly in and out of the replica of the Drake well on the International Exposition grounds Saturday registering their names and experience in the oil industry in the hope that one of the...honors of the exposition would fall to them."[110] When registration ended, 257 pioners had entered their names in the competition and gathered to chat around the monument to be erected in their honor. At the base of this replica of the Drake well was a plaque stating, "This monument when completed is to be dedicated to the memory of those sturdy pioneers both living and dead who by their faithful years of service to the

Vanich Panananda, an official representative of Siam, with President W.G. Skelly in 1938.

petroleum industry have developed it so that we of the living may enjoy its many blessings."[111]

At noon on May 20 at a picnic at the Frank Wolf farm on Sapulpa Road, J.P. Flanagan, acting for Oil Timers committee chairman Charles L. McMahon, announced the winners of the various awards. The title "Grand Old Man of the Oil Industry" went to John W. Van Tine, 81, of Bradford, Pennsylvania, who had worked in the oil industry for 69 years. Accepting the award for him was W.L. Connelly of the Sinclair-Prairie Oil Company. John Edgerton Crosbie, 76, took the title "Grand Old Man of Tulsa." He had lived in the city for 32 years and had worked in the oil industry for 60 years. The awards ceremony concluded with the conferring of the title "Pioneer of Pioneers." This went to M.P. Hanlon, the man who in 1938 could claim the greatest age (88) combined with the greatest number of years of srvice to the industry (78).[112]

The international delegates committee had done its work well in 1938, for W.H. Gardner, chairman of this committee, announced that representatives had come from 26 nations. According to one witness, "By train and by plane Saturday, delegates from every major European nation, from India, Japan and the Straits Settlements, from the republics of Central and South America, men in stiff English shoes and French cutaways will walk Tulsa's streets and the exposition buildings. You will hear a babble of Dutch and Russian, Chinese and Cockney, Swedish, Spanish and Portuguese."[113]

German representatives at the 1938 show. L to R: Hary von Rautenkranz, Karl Grosse, and Werner Mueller. Courtesy Oil and Gas Journal.

An unexpected twist was given to the foreign participation when Maurice Duperrey, president of Rotary International, arrived to recognize the IPE on behalf of his organization. Because of previous commitments, he was unable to appear on the exposition grounds, but he did meet Chairman Gardner at Tulsa Municipal Airport and pose for photographers as Gardner presented him with an international delegate's badge. The audience for this brief ceremony consisted of Madame Duperrey and IPE Manager Way.[114]

There was a full schedule of entertainment for the foreign visitors. This began with a reception and dinner-dance at Southern Hills Country Club with J. Garfield Buell as toastmaster. On May 18 the Tulsa Rotary Club provided a welcoming luncheon for delegates, and the following Friday the Chamber of Commerce sponsored a special luncheon to extend a Tulsa welcome to the foreign guests.[115] This committee likewise arranged activities that were both business and pleasure, such as tours and demonstrations. The foreigners witnessed seismographic demonstrations and took tours to refineries and nearby oil fields. These tours amounted to a mini-overview of the procedures of the American oil industry and helped prepare the visitors for "the greatest industrial exhibit which had ever been assembled in one city."[116]

When the show closed, almost every previous record had been broken for attendance, number of exhibitors, and sales. The show also had become the largest exposition in the world—of any kind—without a midway.[117] With sales of more than $5 million, the IPE had become a positive force for "more work in the Tulsa Territory" and was the leading factor in Tulsa's claim to the title "Oil Capitol of the World." According to the *Tulsa Guide*, it was "the city that 'black gold' built, the Magic City of the southwest where the Indian once pitched his teepee."[118] Also, Tulsa had become the primary center for oil producing, transporting, refining, and marketing concerns. Forty major oil companies and 500 lesser ones and individual operators had headquarters in Tulsa in 1938. Also in Tulsa were many manufacturing concerns, branch factories, and warehouses catering to the oil industry because Tulsa was the purchasing headquarters for the oil industry.[119] Said the *Tulsa Tribune*, the "International Petroleum Exposition is Tulsa's greatest triumph. It is our city's best advertisement. It is our proudest possession."[120]

1940

In May of 1940 hopes were high that the United States might avoid the war already raging in Europe. W.G. Skelly reflected the knowledge that petroleum was a crucial factor in the conflict when he noted, "The product of our industry is the one vital essential to military victory and industrial progress."[121] One IPE particpant that year, Dr. Gustav Egloff, predicted a German collapse and defeat owing to a shortage of oil; according to Egloff, Hitler was "committing suicide" unless he pushed for a speedy end to the conclift, for the German war effort was tied to mobility in machines that required gasoline. Egloff speculated that Germany's war effort consumed 160 million barrels of oil a year, a volume that nation could not sustain. On the other hand, he said, America's petroleum industry, with its ability to produce, would give this country a great advantage in any struggle that might involve it against the German war machine.[122]

In 1940, because this nation was not yet involved in Europe's holocaust, the oil industry could indulge itself in the extravaganza which the IPE had become. Thus preparations began for the 11th International Petroleum Exposition with a strong emphasis on presenting the public with a "true picture" of the oil industry. The goal was to be achieved by distributing the history of petroleum

The IPE grounds in May of 1939. The 12th show was exactly one year away when this photo was taken.

to the media, which had access to the average citizen. The IPE management sent special invitations to leading magazine publishers and representatives of press associations and newsreel companies urging them to come to the exposition for an education about one of the world's critical resources and the industry which sought, found, recovered, processed, and marketed it.[123]

The story to be told was that the oil industry was providing the public with better products at lower cost, that savings achieved by efficient methods and improved equipment were passed on to the public, that few oil men became rich searching for oil, and that the oil industry had steadily improved its safety record.[124] IPE officials felt it was time this story was told, thereby dispelling some of the myths that enshrouded the industry.

Officials of the IPE could look back on 17 years of development which had seen the show grow from "twenty-five exhibitors to more than 800 and from a roped-off street to a 25-acre $750,000 plant,"[125] and these officials anticipated that new records would be set in 1940. President Skelly reported to the *Tulsa World* in September of 1939 that 72 percent of available exhibit space already had been sold and that several exhibitors were considering erecting perment buildings to add to the 25 private and seven IPE buildings already at

An aerial view of the 1940 show. The parked cars surrounding the grounds indicate the crowds attending the exposition. Courtesy Oil and Gas Journal.

the site. Manager W.B. Way reported that the exhibit space already purchased comprised a larger amount than had been sold at any of the nine shows prior to the enlargement of the IPE grounds in 1938.[126]

Most committee chairman for the 1940 IPE had been named by December of 1939. Among those who served were James H. Gardner, president of Gardner Petroleum Company, Tulsa; Barton T. Myers, director of International Petroleum, Limited, Toronto; H.M. Stalcup, vice president of Skelly Oil Company; Emby Kaye, manager of recycling operations at Tidewater Associated Oil Company; J.A. LaFortune, vice president of

Warren Petroleum Company; Dr. Gustav Egloff, director of research for Universal Oil Products Company, Chicago; and R.W. McDowell, vice president in charge of sales, Mid Continent Petroleum Corporation. These chairman quickly mobilized their committees to have everything ready for the exposition, to be held May 18-25. By April 10 the *Tulsa World* could report that the exposition was "in the bag" and that plans already were underway for the 1942 show.[127] Optimistic utterances about the coming exhibition," said the *World*, "and strong views about the twelfth [1942] oil show came from William Grove Skelly, president, William B. Way, general manager and from committee and sub committee chairmen."[128]

At 9:30 a.m. on May 18 the gates of the IPE opened, and "a stream of visitors began pouring into the exposition grounds....Within an hour and ten minutes after the gates were opened, the first official sale had been announced."[129] The huge opening-day crowd seemed determined about viewing the strange and the new, about touching and probing machines, pipes, tubes, and tools, and quizzing booth attendants, greeting friends, and waiting for the official opening ceremonies, which were scheduled for 1:00 p.m. About noon the old nemesis of the IPE came, sending 5000 or more visitors scurrying for cover before the driving rain and necessitating a hurried rescheduling of the opening ceremonies into the "comparatively small quarters of the exhibitors' lounge."[130] The ritual unfurling of the IPE's giant American flag over Drake Drive was changed to Sunday afternoon, but all other ceremonies, although cramped, were held in the exhibitors' lounge, "a room designed to accommodate a dozen or more men who wanted to relax."[131] Members of the American Legion Band, scheduled to hold a brief opening concert, nearly filled the room as they went ahead with a 30-minute performance. Photographers and reporters were crowded into the remaining space, leaving almost no room for the speakers and dignitaries, "but the ceremonies were marked with spirit and enthusiasm."[132]

The principal speaker at this occasion was B.C. Heacock, president of Caterpillar Tractor Company of Peoria, Illinois, who emphasized the spirit of Tulsa and Tulsans as a major contributing factor to the success of the exposition. He spoke glowingly of IPE officials and singled out W.G. Skelly for special recognition, adding that Tulsa's

B.C. Heacock, president of the Catepillar Tractor Company, was the keynote speaker in 1940. Courtesy Oil and Gas Journal.

Chamber of Commerce should be credited for its "alertness" and readiness to accept a challenge. The IPE, he said, was the realization of the dreams of all these civic-minded people who not only could dream but also who could transform dream into reality, adding, "The oil industry is a venture, a venture to put capital to work. Producing oil is one thing, but this exposition represents far more than that. Behind it are the principal reasons for the progress of the industry science, the engineers, chemists and the research men."[133]

Master of Ceremonies E.P. Gilmer introduced other speakers for brief comments or statements of welcome to exposition visitors and guests. One speaker was Tulsa Mayor C.H. Veale; another was Andrew F. Scheoppel, chairman of the Corporation Commission of Kansas; and following him was John Silsbee of Tulsa, who read the names of

W.G. Skelly speaking in 1940. In the foreground, listening intently, is Oklahoma Governor Leon Phillips. Courtesy Oil and Gas Journal.

international delegates who already had registered. Oklahoma Governor Leon C. Phillips was on the program; his comments focused on the independence and self-reliance of the oil industry: "The industry has so far been able to operate under its own steam without help or subsidy from government agencies. It is the responsibility of this industry to produce, protect, preserve that portion of our national resources, and it is a great satisfaction to know that the industry is forward looking and courageous and has not yet given up."[134]

President Skelly took up this theme in commenting on the contribution of the oil industry to the resources of the governmnnt and the public: "This is a business that pays more taxes than any other industry of the United States regardless of size or age. The tax bill of this industry is one and a third billion dollars and that means that the industry pays a hundred dollars a month in taxes on each of its one million employees."[135] Skelly also announced that the theme of the 11th IPE was "progress in spite of international conditions." Reflecting on the war in Europe, he added, "I dedicate this eleventh IPE to the further advancements of the petroleum industry in a world of greater industrial progress, economic prosperity and human happiness. And while the oil industry of America will be ready for any national emergency, let us hope and pray that will never come. We would rather serve a world at peace than at war."[136]

Preparations had been underway since before the end of 1939 to present "another chapter in the progress of the $13 billion [oil] industry in developing more efficient and safer equipment and methods to carry on its operations."[137] A brochure in 1940 proclaimed:

> An oil man can examine 10,000 pieces of oil equipment from manufacturing firms and [visit] 148 cities in twenty-nine states of the United States at the International Petroleum Exposition. It would take months to cover 148 cities and 29 states, yet you can accomplish the same result and see the exhibits from this large territory by attending the International Petroleum Exposition in Tulsa.[138]

These exhibits, according to another report, provided:

> the greatest demonstration in the world of the thousands of pieces of equipment and devices used in [the oil] business....[They] give manufacturers the best opportunity available to become better acquainted with the problems, equipment and necessities of the industry. It is the medium through which contacts may be made to assure the best representation for manufacturers in a market which buys more than one billion dollars' worth of equipment from other industries each year.[139]

One of those "other industries" was automobiles, represented at the 1940 IPE by General Motors. In January it had contracted for more space than any other exhibitor, and it announced plans to erect its own permanent building on the exposition grounds in order to have a convenient forum for displaying GM's complete line of cars, trucks, and engines for oil field work. This was the year of the diesel engine for GM, and a complete display of them was arranged to show their capabilities and economies.[140] Not to be outdone,

Andrew F. Schoeppel, chairman of the Kansas Corporation Commission, speaking in 1940. Courtesy Oil and Gas Journal.

Chrysler Motor Company displayed a "$20,000 working model of one of the world's most modern automobile plants." This model previously had been a popular attraction at the New York World's Fair. Housed in a 25 foot trailer constructed especially for it, the model drew a steady stream of spectators.

Subsidiaries of United States Steel reserved eight spaces in the exhibit area in 1940 to set up a diarama showing how an electric furnace made stainless steel, which was of growing importance in the oil industry.[142]

The oil industry itself was heavily represented in the display area. Reservations for space began as the 1938 show closed; J.J. Tokheim, inventor in 1897 of the glass-topped gasoline pump, had inked a contract for a display area. "I've always figured that it pays to be prompt," said the former hardware store owner. He had started with an investment of $2400 and had built a business worth $100,000 in less than a decade.[143] Tokheim was one of more than 20 exhibitors in 1938 who were so determined to return that they made reservations for space two years early.[144]

Another returning exhibitor was the Franks Manufacturing Company. After generating a flurry of stories in 1938 by striking oil with its "slim hole" rig on the exposition grounds, it announced it would return in 1940 with a portable telescoping derrick mounted on a truck.[145] This unit was capable of servicing a hundred wells, thereby eliminating the need to erect a service derrick above each well. This would end one of the most spectacular sights of oil fields, "a forrest of derricks, hundreds of them, erected almost on top of each other."[146] According to the Franks Company, one of its portable derricks could do the work of 15 to 20 stationary rigs and could provide better service while allowing enormous savings.

One of the most intriguing exhibits in 1940 was the "swampbuggy" exhibited by Amerada Petroleum Corporation, a monstrous vehicle painted a gleaming red and designed for traveling across swamps and marshes. Attendants were kept busy giving visitors rides and demonstrating the unique features of a vehicle part automobile and part truck on a lake created west of the showgrounds. Children and their parents stood in long lines to get a brief ride in the balloon-tired monster.[147]

Another fascinating display was Clark Brothers' eight-cylinder "angle compressor." It was taller than an average man and could develop pressures to 5000 psi with its 800-horsepower engine. Each of its cylinders was "nearly as large as a washtub."[148] Giant machinery also was displayed by International Harvester, a company usually more concerned with agriculture than with the oil fields. In 1940 the firm demonstrated a fleet of tractors suitable for petroleum industry work; these tractors could start operating on gasoline, then switch to heavier fuels when necessary for economy or convenience.[149]

Some visitors in 1940 questioned the housing of millions of dollars worth of equipment in simple corrugated iron buildings. Leslie Brooks later explained that there was a reason for using these huge, rough buildings: "Exhibitors who wanted to install heavier moving equipment could tear up the floor for foundations or punch holes in the walls and roofs for exhaust stacks. Then when the show was over, there was little repair work to do. If we had had brick or concrete walls and heavy

Crowds inspecting Continental Supply Company's giant derrick in 1940. Courtesy Oil and Gas Journal.

IPE visitors were fascinated by giant machinery. Courtesy Oil and Gas Journal.

roofs, a lot of exhibits would never have got inside."[150]

At the Scientific and Technical Building a fire broke out on Wednesday, May 22, in the booth of the Tulsa Camera-Record Company. A disaster was averted when a quick-thinking attendant of the Eugene Dietzgen Company exhibit grabbed a fire extinguisher and smothered the flames before they could spread.[151] There were other fires burning in the technical exhibits, but these were part of the exhibit of the Ozarks Chemical Company of Tulsa, which had set up a large glass windowed tank to show how controlled fire could be used to "revolutionize construction of steam boilers and a host of other deeds for man never dreamed of before."[152]

The star of the technical exhibits in 1940 was "a super-sensitive device for the discovery of new producing horizons in already developed fields [in] producing sands which were passed up in the first drilling operations."[153] Thi͏s had been invented by John T. Hayward of Barnsdall Research Corpora͏tion developed by Engineering L͏aboratories, Tulsa. "What we've done," explaine͏d Hayward, "is put an instrument aboard a ro͏tary that records what formations we're dr͏illing and the potentialities of those for͏mations. With an instrument panel the rig͏ crew can tell at all times how fast the we͏ll is being drilled, the hardness of the formation, an͏d the possibility of washouts and blowouts. The new device made it unnecessary to "run a core," a costly and time-consuming procedure. Only when the core had been tested chemically and a sample log compiled could the operator tell what formations were being encountered. The new device provided this information immediately, accurately, and continuously. It was so new that when it was exhibited it had not yet been named.[155]

Other devices for exploring the subsurface were exhibited in 1940 by Geophysical Research Corporation: the magnetometer and the torsion balance; these relied for their usefulness on differences in gravitational pull in certain types of formations. Such differences were recorded on delicate instruments that allowed geologists to "estimate what is hundreds of feet below."¹⁵⁶ W.G. Skelly's Tulsa radio station KVOO had an exhibit of what was believed to be "the first of the facsimile broadcast senders and receivers ever seen."¹⁵⁷ And an unusual exhibit was provided by Eastman Oilwell Supply Corporation, which specialized in directional drilling. This focused on Eastman's "knucklejoint" or "whipstock" which made it possible to drill a well in any chosen direction from a derrick. This device also made possible the drilling of several wells from the same platform to tap oil pools great distances apart.¹⁵⁸

In addition to the samples of equipment and the instruments in the various exhibits, the Hall of Science was crowded with models of oil field operations. The Hall in fact was a haven for the model builder "whether he preferred moving or stationary objects."¹⁵⁹ There were models of steam and gasoline drilling rigs, gasoline plants, automobile factories, tractors, and other vehicles, even a model of an entire oil field turned "wrongside out."¹⁶⁰ This display, provided by the University of Oklahoma, showed "instead of holes, slender columns corresponding to the wells. At appropriate places on the columns [were] markers indicating the various formations passed through by the drill bit."¹⁶¹ Visitors circled the exhibit to bend over, peer in, and look around it at great length.

Also in the Hall of Science was an exhibit showing the technical progress of safety procedures throughout the development of the oil industry. Overseeing it was O.B. Badger, who had charge of adult education for the Tulsa Public Schools.¹⁶² And on the grounds of the IPE, visitors could see demonstrations of the latest safety measures and equipment being carried out by teams competing for prizes and honors. The safety committee, headed by R.S. Huffman, personnel and safety director for Oklahoma Natural Gas Company, had chosen as its slogan, "Let's Make Petroleum the Safest Industry in the World."¹⁶³ Assisting Huffman were R.J. Lauder, Shell Oil Company; J.H. Savage, Mid Continent Petroleum Corporation; L.G. Clark, Gulf Oil Corporation; J.A. Fleming, Indian Territory Illuminating Oil Company; C.V. Williams, Employers' Liability Assurance Company; and R.W. Cooper, Oklahoma Power and Water Company. Others involved in planning and directing the safety events were H.W. Boggess, superintendent of safety for

Booth of the California Oil World in 1940. Later this became the Pacific Oil World. Courtesy Pacific Oil World.

Sinclair-Prairie Oil Company and chairman of the central committee of the Mid Continent Oil and Gas Association, and J.C. Young of Tidewater Associated Oil Company.¹⁶⁴

Contests were held in each of the four major branches of the oil industry: producing, refining, piepining, and marketing. In each division the superintendent with the longest record of accident-free man-hours worked was presented a wristwatch by the IPE and entertained at a banquet. Watches also were given to the drivers of oil-company cars and trucks with the most accident-free miles driven. On Thursday, May 23, an all-day safety conference was held to emphasize the importance of safety in oil field operations. This was attended by many visiting editors of trade journals and oil house organs and journalists accorded the status of special guests of the IPE.¹⁶⁵ These newsmen were provided with a special office to help them gather any information they wanted, and they had access to a photographer who would take any picture needed to accompany a news release. Leslie Brooks, IPE publicity director, appointed Luther Williams of the Mid Continent Petroleum Corporation to head a committee to make arrangements for entertaining the editors. One event he planned was a one-day session of the Southwestern Association of Industrial Editors, of which Williams was president. All visiting editors were special guests at this conference, and a huge party later was held for them at the Tulsa Club.[166] The writers and editors responded to the hospitality of the IPE by giving it wide coverage in their journals.

A favorite event was the Old Timers contest because it made interesting newscopy. In 1940 there was a major change in this feature. Previously, awards had gone to the Grand Old Man of Tulsa, the Grand Old Man of the Oil Industry, and the Pioneer of Pioneers. In the 11th IPE, the number of titles was increased to 10. Awards were given in the areas of production, pipeline and transportation, refining, marketing, supplies and equipment, natural gasoline, natural gas, purchasing agents, and land-lease and legal. These awards were made "on the basis of service and of contributions to the industry, regardless of division," and all nominations for a particular award automatically included a nomination for the Pioneer of Pioneers award.[167]

Those chosen for these honors included George E. Tabor, refining; William M. Schermerhorn, marketing; Forrest Towl, pipeline and transportation; Howell C. Cooper, natural gas; Frank L. Chase, natural gasoline; James A. Veasey, land-lease and legal; John M. Crawford, supplies and equipment; Guy T. Berry, production, and Frank D. Bryant, purchasing agent. P.H. "Patsy" Mack, 88, of Bradford, Pennsylvania, was named "Pioneer of Pioneers."[168] Other Old Timers, runners-up to the winners, were given "Officiates of Service" by the Old Timers memorial committee, which was chaired by C.L. McMahon. The awards were made on Drake Avenue near the replica of Colonel Drake's wooden derrick.[169]

Youngsters of the oil industry likewise were recognized during the 11th IPE in a special program hosted by the Tulsa Jaycees on May 22, "Junior Chamber Day." W.A. Baden chaired this committee with C.M. Hardy and Frank Mattoon assisting. Young oil executives were brought together at a luncheon in the Tulsa Building, and from there they went on an oil field tour to observe on-site operations.

The engineering school of the University of Tulsa invited students from the automotive and machine shop departments at Kansas State Teachers College at Pittsburgh, Kansas, to visit the exposition. These students were given free admission tickets by the IPE. Invitations also went to students at the Colorado School of Mines, the University of Kansas, Oklahoma A. & M. College, the University of Texas, Kansas State College, the University of Arkansas, and the University of Oklahoma. Reprsentatives attended from all these institutions.[170]

Yet another group was coordinating the visit of foreign delegates to the IPE. The NOMADS Club, formed by those who handled foreign sales for the various oil companies, held its first meeting on January 8, 1940, and elected Fred E. Cooper, of Fred E. Cooper Company, president; H.M. Cosgrove, of the Purchasing Agents Association, secretary; William G. Green, of Seismograph Service Corporation, to the by-laws committee; and C.O. Willson, of the *Oil and Gas Journal*, also to the by laws committee.[171] Soon after assuming responsibility for foreign visitors to the IPE, the NOMADS announced their intent to erect a new building on the exposition grounds to serve as headquarters for all activities involving foreign visitors. Thanks to the beneficence of W.R. Braden, of Braden Steel Company, this building was erected at no cost to NOMADS or the IPE. Braden, at the time chairman of the exposition committee of NOMADS, was keenly aware of the need of such a building. Construction began during exposition week after an agreement was reached to locate the structure at the right side of the main entrance to the grounds.[172]

Foreign oil interests, particularly from Europe, were strongly represented at the 1940 IPE because everyone was aware that modern warfare required huge amounts of oil. Leslie Brooks, in an article for the *Tulsa World* following his trip to visit embassies in Washington, D.C., accurately pointed out, "The next war will be fought with oil and the diplomats in Washington…don't mind letting you know that with conditions like they are in Europe that they are quite interested in oil and they believe that Americans, or rather United States oil men, know where and how to find oil."[173]

An interesting sidelight on the war-heightened interest of foreign delegates to the IPE was the visit of three Japanese representatives of Nippon Oil Companay of Tokyo. They were in the United States to tour American oil fields in search of the latest developments in equipment and technology which they needed for the war that would begin in December of 1941.

Like the 10 IPE shows that had preceded it, the 11th exposition closed with an outpouring of

Aerial view of the 1940 IPE with the downtown skyline in the distance. The barren area at the bottom (east of 21st and Yale) is now fully developed. In the 1940s the IPE grounds were on the eastern outskirts of Tulsa. Courtesy Howard Hopkins.

praise from its many constituencies. The *Tulsa World* called it the "Greatest Oil Expo."[171] The enthusiasm generated carried over into immediate planning for the 1942 show, planning which began even before the exhibits were cleared from the grounds in 1940. What no one could have known at that time, not even the three Japanese representatives of Nippon Oil Company, was that the 12th IPE was a world war and eight years away.

CHAPTER THREE
BACK TO BUSINESS
(1948-1959)

The petroleum industry, perhaps more than any other segment of the economy in the aftermath of war, needed time to redirect its priorities and resources. Consequently the IPE did not reopen until 1948, but when it did resume it showed no signs of having lost its appeal or its stature in the petroleum industry. Delegates, exhibitors, dignitaries, and sightseers again flocked to Tulsa for the great show, thousands of them to examine, learn, buy, or sell, and thousands more simply to look. What they found was an industry revitalized by the developments and advances that had taken place since 1940.

The show in 1953 opened the fourth decade of the IPE's existence, during which time it had become the world's leading industrial exhibit. It would continue to showcase the latest and best petroleum equipment, supplies, technology, and ideas and would fulfill its dual goal of displaying and explaining the state of the petroleum art and of predicting improvements soon to come. Members of the oil fraternity had learned to expect an array of miracles when they convened in Tulsa, and they were not disappointed.

When they came to Tulsa in 1959 to celebrate the 100th anniversary of their industry, they found it still booming and almost reborn by a new burst of development and advances which had occurred in just six years. The IPE had long since become a "required course" for oil entrepreneurs who wanted to stay at the forefront of their industry. Yet tradition had not been forgotten. If Colonel Edwin Drake, founder of the oil industry, had been there in 1959, he would have found enough of his tools and derrick to gnaw another 69½-foot hole.

1948

Exploratory drilling reached an all-time high in 1948, and that year the greatest quantity of new reserves since 1937 were discovered. During the previous year almost 11,000 wildcat tests had been drilled, and thanks to improved prices the production of oil and raw condensates reached a record-breaking 1.86 billion barrels. More important, 1.39 billion barrels had been replaced in the underground inventory, bringing the nation's known reserves to 23.34 billion barrels as of January 1, 1948. Every projection for the year ahead indicated that the demand for refinery production would exceed all previous records. The only limiting factor foreseen for 1948 involved distribution facilities and adequate equipment and supplies, especially of steel. Yet this was seen as a challenge as the nation's second largest business geared up for the 12th International Petroleum Exposition and Congress.[1]

By late 1947 the oil industry had resumed business-as-usual although there were rumors that the IPE was through. General Manager William B. Way met this rumor head-on at a press conference on November 15, 1947, when he declared, "I don't know who starts such rumors or where or why....All reports of no more oil expositions in Tulsa are unfounded."[2] The dates for the 12th IPE were May 15-22, 1948, and work immediately began to prepare the exposition grounds and to organize the various committees needed to coordinate the mammoth show.

In April of 1947 the Tulsa County Fair Board had approved the application of the IPE for an additional plot of ground 100 feet wide extending 900 feet east and west along the north side of the Texas and California buildings. This space, the IPE said in its application, was needed to provide for the anticipated expansion of the show. By that date requests for space already were greater than for any previous show, and it was expected that the 12th show would be the largest ever. By July of 1947 plans for improvements of the new space and for other areas on the grounds had been submitted by the Shibley Engineering Company of Tulsa and had been approved by the IPE executive committee, which consisted of W.G. Skelly, C.H. Pape, Alf G. Heggem, H.R. Powers, E.F. Bullad, W.M. Bovaird, Frank J. Hinderliter, and

William B. Way. Hinderliter and Way had been appointed as a special grounds committee and had been authorized to obtain bids for installing gas, sewer, and power lines as well as for a 20-foot paved road linking Drake Drive on the south with Skelly Drive on the north.[3]

This new street was to be a "White Way of dazzling brilliance" and was to be occupied by exhibitors such as Bethlehem Supply Company of Tulsa, General Motors Corporation of Detroit, the George E. Failing Company of Enid, the Brewster Company of Shreveport, Louisiana, and Dresser Industries of Cleveland. Their combined space requirements amounted to more than 60,000 square feet. This "strong support" by industrial giants, said Manager Way, "make the new improvements possible and still further assure the permanency of the International Petroleum Exposition as one of Tulsa's major industrial institutions."[4]

Improvements in communications and coordination also were made for the 12th show, primarily through opening a convenient downtown office in the French Room of the Hotel Tulsa where calls could be taken and reservations made. According to W.G. Skelly, the opening of this office spurred an avalanche of calls: "As soon as oil men heard we were contemplating opening a convenient downtown office, they began to call in person, telephone and wire us to double, treble and in many cases quadruple their previous contracts for space."[5] When Skelly made this statement, the 12th IPE was still 11 months away, and already 90 percent of available space was under contract. The eight-year closing of the exposition had piqued the interest of industry leaders in resuming their most successful joint venture, for the period 1940-1948 had been some of the most inventive years of the oil industry. The 1948 show promised to contain "more that is new...than ever before, as well as more men in the industry who should see everything..., and eight days is such a short time."[6] *DuPont Magazine* in April of 1948 noted, "Next May during a period of eight days, Tulsa, Oklahoma, will play host to about 250,000 visitors..., and oil men the world over will gather to see and hear what's new in America's second largest industry."[7]

The duties of hosting in 1948 were assumed by eight industry leaders, each man serving as host for one day of the exposition. These men included Howard L. Berkey of the Tret-O-Lite Company; J.H. Brooks, Republic Supply Company; W.L. James, Stanolind Oil and Gas Company; J.A. LaFortune, Warren Petroleum Company; R.W. McDowell, Mid Continent Petroleum Corporation; J.R. McWilliams, Carter Oil Company; Mark S. Patton, Hurley Gasoline Sales Company; and J.L. Shakely, Jones and Laughlin Supply Company. This group met during the week of May 4 to discuss their duties and to select other Tulsans to assist them.[8]

A major concern for the 1948 IPE was providing lodging and food for the 20,000 visitors from out of town expected for each of the eight days of the show. Nine months before the IPE opened, the housing committee mobilized under civic leaders R.L. Ledterman and C.C. Brann to make plans for "squeezing an anticipated 20,000 guests into 1500 hotel and motel rooms plus accommodations that must be unearthed somewhere."[9] Ledterman and Brann coordinated their work with people like Frank Bentley of the Mayo Hotel and Hollis Hodges of Hotel Tulsa, both of whom reported that requests for lodging had begun pouring in even before the dates for the IPE had been announced. Manager Way reported, "They [oil company representatives] have even gone so far as to convass hotels in nearby towns and the better class tourist courts. One thing is certain, the exhibitors themselves would take every room in the city if allowed. But that would leave no room for the visitors who are the customers of the exhibitors. Without customers the exhibitors would have no reason to come here."[10]

The basic policy adopted by the committee was to refuse hotel rooms for purposes of entertainment and to require that all requests for rooms received by individual hotels be forwarded to the IPE housing committee, which would have final authority to allocate space. The committee solicited and received pledges from all major hotel managers to reserve their entire facilities for IPE visitors. However, hotels and motels could not provide all the lodging that would be required for IPE visitors and participants, and on March 9 the call went out for Tulsa citizens to make room in their homes for the overflow. The *Tulsa Tribune* estimated that 5000 rooms would be needed and challenged Tulsans: "If we cannot accommodate [IPE visitors] it will prove that Tulsa is too small a city for an exposition of this size....Here's a civic

Aerial view of the 1948 IPE. Parking space obviously was at a premium. Courtesy Petroleo Interamericano.

job for everyone with a spare room. Every Tulsan is interested as a matter of pride in putting over the Petroleum Exposition....Let's get those 5000 rooms."[11] Tulsans responded with a flood of calls to the IPE office, proving once again the city's pride and involvement in the IPE.

The 12th IPE opened at 10:00 a.m. on the morning of May 15 at the NOMADS International Headquarters building on the northwest corner of the exposition grounds. The *Tulsa World* described this as the coronation of an "industrial giant" at facilities that were "a castle built by free enterprise."[12] Headlining that first day's program was Secretary of the Interior J.A. Krug. He told the assembled oil men that their industry had more reason for optimism than at any time since World War II. "This show," he said, "is a legitimate mark of optimism for the future of the oil industry," optimism based on: "1) discovery of proven reserves totalling 21 billion barrels, the highest in history; 2) an increase to two billion barrels a year in do-

W.G. Skelly addressing the crowd in 1948 from the speaker's platform atop the NOMADS Building. Courtesy Oil and Gas Journal.

mestic crude production, also the highest in history, and 3) increased demand for petroleum, reaching 2 billion, 150 million barrels in 1947 and probably more this year."[14] Krug also cited oil industry assets of $22 billion, the employment of 1.25 million people, and the drilling of 40,000 wells as proof that the industry was among the top businesses in American and was advancing steadily.

Other opening-day speakers included Oklahoma Governor Roy J. Turner; Tulsa Mayor Roy Lundy; W.R. Boyd, Jr., president of the American Petroleum Institute; Dr. Jose Martoreno Battisti, president of the Petroleum Bureau of Venezuela; and Clark Hungerford, president of the Frisco Railroad. Also participating in the opening ceremonies was a delegation of 20 oil men from Venezuela sent by Minister Perez Alfonso. Each speaker praised the industry for its advances and predicted yet more progress in the future.[15]

Following the opening ceremonies the crowd dispersed into the 4.5 miles of exhibits, but they reassembled when fireworks exploded overhead to signal the unfurling between two oil derricks of the IPE's giant American flag. It was hoisted "as a symbol of American free enterprise and the system under which Tulsa has taken the lead in the petroleum industry."[16] As this great banner unfurled, more than 5000 miniature flags fluttered from it down to the waiting hands of the crowd. These tiny flags were suddenly flooded with light and the grounds resounded to the whir and grind of machinery coming to life for the eight-day run of the IPE.

In 1948 the IPE, with its displays of equipment and machinery, shared the limelight with a revitalized International Petroleum Congress, which was chaired by Alf G. Heggem. His leadership, thought IPE officials, would make this congress "the most important such meeting since 1937 when a World Petroleum Congress had convened in Paris."[17] In addition to being the first person ever to bear the academic title "petroleum engineer," Heggem was a past president of the Tulsa Chamber of Commerce and an energetic civic leader as well as an IPE vice president. All meetings of the oil congress were of the open forum type so that speakers would appeal to all interest-groups. Participation was welcomed from governmental agencies, equipment, supply and engineering interests, as well as the oil industry in general, both foreign and domestic.[18]

The 1948 congress included panels discussing "current oil research, future prospects, and problems facing industry on an international scale."[19] Government officials and oil industry representatives from more than 30 countries attended. The affair opened on May 18 with an address by Governor Beauford Jester of Texas, who was chosen, Heggem said, because he was "head of the world's largest geo-political unit in crude oil production" and because of his leadership in the fight for state ownership of the Tidelands."[20] Other speakers at the congress included N.E. Tanner, Canadian Minister of Lands and Mines; Antonio J. Mermudez, general manager of Petroleos Mexicanos (the Mexican government's oil agency); Benjamin

An informal gathering of IPE officials and guest speakers in 1948. L to R: American Petroleum Institute President William R. Boyd, Jr., William B. Way, Secretary of the Interior J.A. Krug, Oklahoma Governor Roy J. Turner, W.K. Warren, W.G. Skelly, and Jose Martorano, director of Venezuela's petroleum board. Courtesy Oil and Gas Journal.

Scale working model of a rotary rig. This was built and used by Kilgore (Texas) Junior College in courses sponsored there by the American Association of Oilwell Drilling Contractors. Courtesy Oil and Gas Journal.

Farrless, president of United States Steel Corporation; Charles Kettering of General Motors Corporation; and Dr. Gustav Egloff of Universal Oil Products Company.

The Congress met on May 18, 20, and 21 in the north and south auditoriums of Tulsa Central High School and was open to any member of the oil industry who wanted to attend. Abstracts of all papers presented during the meetings were published and available to participants. Translators also were available to visitors not familiar with English. Oil ideologies were heard ranging from strict national control to unlimited free enterprise. H.B. Fell, vice president of the Independent Petroleum Association of America, spoke passionately of the progress made by the industry under the competitive system of free enterprise in the United States, the congress itself one of the main products of that freedom.[22]

Securing international participation in the exposition and congress had been the job of Dr. Oscar B. Irizarry, named the IPE's ambassador at-large to a dozen Central and South American nations. Dr. Irizarry, editor of *Petroleos Interamericano*, the Latin American affiliate of the *Oil and Gas Journal*, had been loaned to the IPE, as President Skelly explained,

> at the suggestion of our general manager William B. Way and myself. We are grateful indeed to Mr. [P.C.] Lauinger for his cooperation. Dr. Irizarry's assignment is one of the most important steps ever taken in building acceptance for and attendance from petroleum personnel in South American countries. His work will supplement that of IPE Director Robert C. Sharp, independent oil operator of Tulsa who is presently in South America.[23]

During this three months of travel, Dr. Irizarry carried special invitations to the IPE from President Skelly and Oklahoma Governor Roy J. Turner. His itinerary of some 20,000 air miles included Havana, Cuba; Caracas, Venezuela; Trinidad, British West Indies; Rio de Janero and Sao Paulo, Brazil; Montevideo, Uruguay; Buenos Aires, Argentina; Santiago, Chile; Lima and Talara, Peru; La Paz, Bolivia; Quito, Ecuador; Bogota and Medellin, Colombia; and Balboa and Panama City, Panama. In each country Dr. Irizarry presented hand-engraved invitations to prominent government officials and oil industry leaders.

The foreign delegates and visitors who came to Tulsa found a full schedule of professional and social events planned for them, all coordinated and hosted by NOMADS. Chapters of this society from Dallas, Houston, New York City, Los Angeles, and Tulsa cooperated in planning entertainment and accommodations for foreign oil men. Committees chaired by Tulsa NOMADS were mobilized to make arrangements for these visitors. The finance committee was headed by J.M. Walker of the Lane Wells Company; the treasurers of each chapter of the organization made up the rest of the committee. Housing and registration were directed by Dr. Oscar Irizarry, assisted by H.M. Cosgrove, executive secretary of the Na-

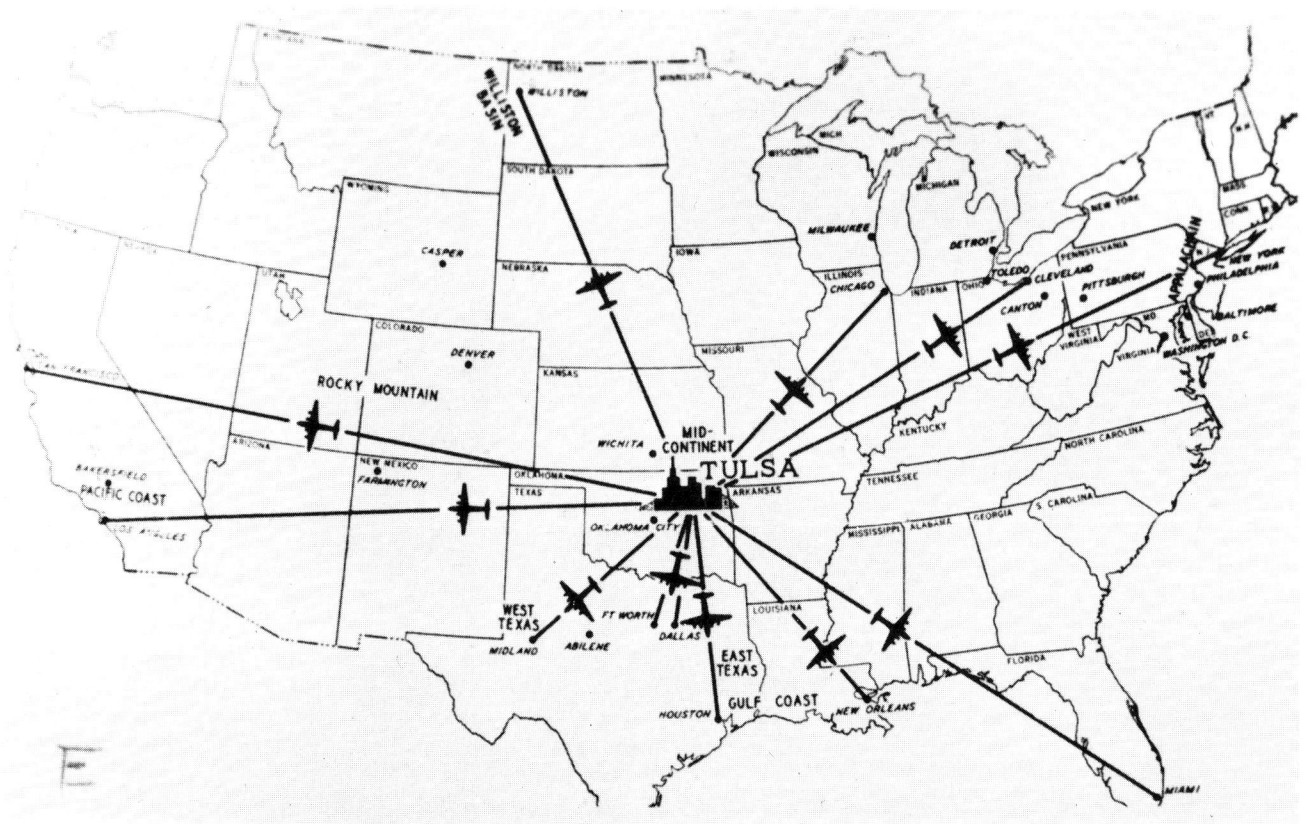

Flying time to Tulsa from various points of entry from foreign countries. Foreign participation was emphasized in 1948. Courtesy Petroleo Interamericano.

tional Board of Regents of NOMADS; Shep Miers of Southwest Supply Company; and John G. Staudt of Dowell, Incorporated. Cosgrove also handled NOMADS publicity, and H.M. Cooley of Bethlehem Steel Company headed the NOMADS international housing committee assisted by one member of each NOMADS chapter. This committee kept a member on duty at the NOMAD International House every hour the show was open. NOMADS members also were available thoughout the show "to help in any way and to give every possible assistance to the hundreds of delegates, government officials and oil company personnel from abroad."[24] Maintaining the headquarters was in the hands of Roy R. Bush of Pittsburgh Equitable Meter Division of Rockwell Manufacturing Company, Oppie Dimmick of Century Geophysical Corporation, and H.M. Cosgrove.[25]

While NOMADS was hosting the international segment of the exposition, the Tulsa Chapter of the American Society of Safety Engineers was planning the "Safety and Oil" segment of the IPE program. This society had a variety of events and demonstrations to show IPE visitors that "the petroleum industry which was once regarded as hazardous is today one of the safest industries."[26] The latest equipment and techniques for protecting oil personnel in the field and for administering life-saving first-aid treatment were demonstrated by members of the society. Many of the techniques they demonstrated were useful not only in the oil fields but also on the exposition grounds where as many as 100 first-aid cases a day resulted from accidents and sudden illnesses. In fact, the call for emergency medical services on the grounds had become so great by 1948 that a four-room hospital was installed in the original NOMADS headquarters on the northwest section of the grounds. This hospital was staffed by six members of the U.S. Cadet Nurse Corps from St. John's Hospital. According to Manager Way, these facilities compared favorably with those of larger insitutions offering a wider range of medical services.[27]

The Society of Safety Engineers also joined with the American Trucking Association to present a "road-e-o" at the 1948 IPE show. This event featured [...] hauling equipment [...] "road" [...] unload- [...] de- [...] kwood [...] Drake [...] the ex- [...] ere as- [...] likely [...] drivers [...] ere stu- [...] m engi- [...] student [...] chanical [...] the uni- [...] as, and [...] d Okla- [...] ese stu- [...] III sub- [...] of ASME [...] Gulf Oil [...] nsulting [...] nt branch [...] y student [...] uctor and [...] s possible

Many college instructors who've written our

offices don't yet know the big educational bonus planned for them by the IPE scientific and technical committee. [The Hall of Science] will be the most comprehensive museum of the oil industry ever assembled. It will house many visual and dramatic petroleum presentations from all over the world. All of the exhibits will show how indispensable oil has become in our life and time.[30]

At the opposite end of the age scale were the men concluding their active careers in the oil industry, the Old Timers. More than 600 men with more than 25 years of service registered for the competition in 1948, and each received silver "pioneer" buttons. Eight of the 600 were honored for having the greatest length of service and and contributing the most to their branches of the industry.[31]

The "Pioneer of Refining" was Charles L. Suhr of Oil City, Pennsylvania, chairman of the board of Pennzoil Company with almost 60 years of service to the industry. In the pipeline division, B.E. Hull of Houston, president of Trans-American Pipeline Company, vice president of Texaco, and a director of five other pipeline companies, was honored for 45 years of srvice. The pioneer in natural gas was Ralph W. Gallagher of New York, the retired board chairman of Standard Oil of New Jersey. The pioneer in land-lease and legal went to Dana H. Kelsey of Tulsa, vice president of Sinclair-Prairie Oil Company and a veteran of 43 years of service. George W. Bovaird, a 54 year veteran of the oil industry from Bradford, Pennsyvlania, received the pioneer award in supplies and equipment. In the production division the pioneer was Herbert Straight of Bartlesville, Oklahoma; he was chairman of the board of Cities Service Oil Company and had 51 years in the oil industry. The coveted award "Grand Old Man of Oklahoma" went to Frank Phillips of Bartlesville, chairman of the board of Phillips Petroleum Company and 45-year veteran of the industry.[32]

The premier honor among Old Timers went to Frank A. Haskell of Sharon, Connecticut, who was named "Pioneer of Pioneers" by the Old Timers Committee, which consisted of W.L. Connelly, chairman, James A. Veasey, J.K. McGoldrick, James P. Flanagan, and E.B. Reeser.[33] Haskell was a retired vice president of Tidewater Oil Company but still a director of the firm. He had started in the famous "Pitt Hole" in Pennsylvania where his father had been an operator during the boom of the 1870s, then had moved to Oklahoma in 1911 as head of the Oklahoma Oil Company (later absorbed into Tidewater). Haskell had 69 years of service in the oil industry. He, along with the others honored, was recognized and given an elegant commemorative medal at a ceremony near the replica of the Drake Well.[34]

History was made on the exhibit grounds in 1948. More than $100 million in tools, equipment, and supplies were on display in the industry's showcase. One focus of interest was the Reynolds Metals Company's aluminum bus especially equipped to demonstrate the usefulness of aluminum to the oil industry. Among the many uses for aluminum shown were welded pipe fittings, special couplings of several types, samples of coated pipe for underground use, threaded pipe for shot-hole casings, and other specialty items. The display coach itself demonstrated what aluminum could mean in portable equipment.[35]

In 1948 the worldwide oil industry was composed of some 34,000 firms, most of which were represented in some way at the IPE. Many were conspicuous by their giant displays; for example, Dresser Industries of Cleveland, Ohio, had 17,703 square feet of display space. International Harvester was showing a 23-ton crawler-tractor, the largest and most powerful in the world. Halliburton Oil Well Cementing Company had an exhibit that featured models, animated shadow boxes, subtle color combinations, and special lighting effects to guide visitors through the exhibit, all of it set off by a huge globe spinning in the center of the Halliburton Building and pinpointed with lights to show Halliburton bases around the world.[36]

The DuPont Company's display in the Texas Building was constructed around a design featuring DuPont Lucite and a variety of plastics. Outside, the Continental Supply Company, the exposition's second-largest exhibitor, had a radio station built in the midst of the exhibits; FM communications were carried on from a 30-foot antenna atop Continental's 146-foot derrick (one of two derricks installed on the grounds by Continental). Inside the Continental Building were exhibits that included a miniature standing mill used to manufacture wire rope and a miniature pipe mill in full operation. The total value of the Continental exhibit was estimated at more than $3 million.[37]

National Supply Company's exhibit was made up of 40 carloads of equipment valued at more than $1 million. Included in it were three turbo supercharged dual-fuel diesel engines which worked on gas or oil and which had a combined horsepower of 2010 that drove National's big 160-foot rig which was designed to drill to 20,000 feet with 4.5-inch drill pipe; this used an E-700 pump, the world's largest mast pump at 700 hydraulic

Technical exhibits in 1948 showing a walking beam (left) and the new "free pump" (right). Courtesy Oil and Gas Journal.

A crowd studying the upper and lower sections of a new centrifugal compressor at the 1948 IPE.

horsepower output. Another supply company, Mid Continental, had erected a 63-ton derrick which it exhibited in full operation.[38]

Cooper-Bessemer Corporation brought the world's largest natural gas diesel compresser which, when stripped for shipment, still weighed 120 tons and needed five flat cars for transport. Only seven of these giant compressors had been built. Cooper-Bessemer's compressor was the single largest piece of equipment at the 1948 exposition.[39]

The Franks Manufacturing Company announced at the show a contest in which a new Buick would be awarded to the drilling or production man who completed the jingle best describing Franks' oil field parts and repair service. According to Carl F. White, president of Franks, the purpose of the contest was to focus attention on the scope and quality of Franks' recently expanded and improved field service from Tulsa; Kilgore, Odessa, and Alice, Texas; Great Bend, Kansas; and Compton, California. The Buick went to Perry L. Nunley of Pennwell, Texas, a district foreman for Skelly Oil Company.[40]

Exhibits with a different purpose were featured at the Hall of Science. These, according to Hall of Science committee chairman Dr. B.B. Weatherby, were designed to give "easily understood answers and explanations to questions about the discovery of oil, its production, its transportation, its manufacture, its marketing, its application to the needs of people, and its contributions to their welfare."[41] The most dramatic of these exhibits was one prepared by the American Chemical Society to demonstrate the potential of nuclear power for peace or war. Designed by the editorial boards of two American Chemical Society jour-

President Skelly (center) awarding a new Buick to Perry L. Nunley (right) on behalf of the Franks Manufacturing Company. Looking on are officials of the Franks Company. Courtesy Getty Refining and Marketing Company.

nals, *Industrial and Engineering Chemistry* and *Chemical and Engineering News*, this 13-panel exhibit had been shown in Paris at the request of the United Nations Educational, Scientific and Cultural Organization. A pattern of lights in one panel depicted the shattering of an atom's nucleus by bombardment with a neutron and the resulting scattering of fragments, including other neutrons which in turn split the nuclei of other atoms until a myriad of atoms exploded. The differences in peaceful and destructive uses of atomic power were shown in contrasting photographs that showed the Nagasaki blast as compared to the contributions which nuclear energy could make to science and industry.[42]

Other displays included geologic maps, cross-sections, and aerial and subsurface surveys; these explained how geologists worked, as did seismographs, gravimeters, and magnetometers. There also were maps, charts, and diagrams to show how oil was being hunted below the ocean, and these

One of the biggest oil valves in the industry in 1948. Approximately 2500 tons of catalyst would flow through this valve per hour. Courtesy Oil and Gas Journal.

were supplemented by a 3600-pound diving bell used by R.H. Ray in taking gravimetric readings on the floor of the Gulf of Mexico. Explaining all this was geologist K.K. Kimball.[43]

There were displays to show how drilling had been done in the past, how it was being done at present, and how most engineers thought it would be done in the future. These displays were provided by the Hughes Tool Company, Eastman Oilwell Surveying Company, Halliburton Oilwell Cementing Company, Dowell, Johnson and Fagg Engineering Company, Koge, Reda Pump Company, Tulsa Testing Laboratories, and Core Laboratories. Yet other displays showed the methods of extraction, pilot plants, refining processes perfected since World War II, and myriad other details of the oil industry of interest to both laymen and scientists.[44]

According to the *Tulsa World*, the 1948 IPE surpassed all previous shows. "Past performance meant little" it said, as the present show "broke all records of any Tulsa event of similar duration with an attendance of 300,332, which eclipsed by nearly 75,000 the previous attendance record."[45] Authoritative estimates placed sales at some $1 billion. Thus it was with considerable pride that President W.G. Skelly could say, when the gates closed on Saturday night, that the 1948 IPE

> ...has accocmplished two objectives of worldwide significance. It has proved that the ingenuity, resourcefulness and cooperation of

the petroleum industry will continue to meet the demands for petroleum products in a world of peaceful progress....The exposition and congress also has done an extraordinary job of public relations on an international scale. Representatives from some forty nations have been impressed with the fact that the United States both seeks and warrants the confidence and friendship of peoples everywhere....The exposition is an amazing demonstration of the free-enterprise system in a democracy. It has shown and told the world that private initiative and the ability to work, dream, develop and accomplish things is neither handicapped nor penalized by a democratic form of government.[16]

1953

In 1952 the IPE again was a casualty of war. In 1951 the show's executive committee voted to cancel the exposition because of the Korean Conflict. In a statement issued jointly by General Manager W.B. Way and President W.G. Skelly, the executive committee explained that it had decided to postpone the show for a year owing to "the current critical materials and manpower situation."[47] The *Tulsa World* added, "The schedule of one big show every four years is now disarranged by world affairs...[and] the mighty petroleum industry and the City of Tulsa are convinced of the necessity of postponement and are confidently preparing for the world showing in the spring of 1953."[48]

Among these preparations was the addition of some 35,000 square feet of outside exhibit space thanks to an agreement with the Tulsa Exposition and Fair Corporation. This space was added west of the exposition grounds and south of the fair's poultry building (previously used for Hall of Science exhibits). This area was leveled and graded, and water and gas lines were installed, as also were drainage facilities, roads, and walks.[49]

Housing for the thousands of out-of-town oil men expected to attend the 1953 show was another great challenge. The drive to locate space was led by John B. Marshall of the Tulsa Chamber of Commerce. Barton Myers, Chamber Director, noted that Marshall had put together "the best housing program in the history of the exposition."[50] By December of 1953 Marshall had more

A forest of derricks at the 1948 IPE. These began appearing as portable rigs became available. Courtesy Howard Hopkins.

than 1150 private homes listed for rent to oil show visitors.⁵¹

Although heroic efforts by the housing committee always had produced adequate housing for participants and visitors, considerable thought was given to finding a way to ease this problem, which was increasing with each show. One proposal long under consideration was to increase the length of the show in order to allow companies to send half their personnel to the show, then have them go home so the other half could attend. Such an arrangement not only would ease pressure on housing, but also it would allow more oil people to visit the show and thus make exhibits more cost-effective. At the end of the show in 1948, a survey had been done by mail to ask if participants and exhibitors would support a longer show. More than half of the exhibitors had responded enthusiastically, and the IPE executive committee responded by settling on a 10-day run in 1953—from May 14 to 23 inclusive.⁵²

Within days of this announcement, requests for exhibit space poured into IPE headquarters. By September of 1952 more than 500 requests had been received, leading to frenzied work to repair, remodel, and expand facilities at the exposition grounds.⁵³

In August of 1952 Governor Johnston Murray sent Shirley Barbour, Miss Oklahoma of 1952, on a 16-day tour of Eastern cities as "goodwill ambassador" to promote the IPE. She represented the governor, who at the time was touring South America to present foreign oil executives with a special invitation to the exposition. Miss Barbour carried invitations bearing the governor's seal and signature to deliver to dignitaries she met on her trip. This tie-in between the IPE and the Miss Oklahoma Pageant was conceived by John B. Marshall, Pageant director and manager of conventions for the Chamber of Commerce. Accompanying Miss Barbour were Marshall, Mrs.

Marshall, and Leslie Brooks, IPE publicity director.⁵⁴

One of the best promotional events for the IPE occurred in the oil patch itself when on May 6, 1953, diamond bit Number 29 broke the four mile-deep barrier at Ohio Oil Company's KCL "A" 72-4 in Kern County, California.⁵⁵ This was on every oil man's lips when the exposition gates opened on May 14 to the blare and crescendo of five huge bands, the roar of heavy machinery, and the stir and murmur of thousands of oil men from around the world. Most of the spectators quickly gathered around a platform in front of the NOMADS Building on Drake Drive and Silver Lane to await the raising of the IPE's giant American flag.⁵⁶

On the platform were Charlton H. Lyons, president of the Independent Petroleum Association of America and keynote speaker; Oklahoma Governor Johnston Murray; Tulsa Mayor C.M. Warren; IPE Manager W.B. Way; IPE President W.G. Skelly; and other officials and dignitaries. Lyons' keynote speech focused on incentives for productivity provided by a free economy and "diversified private ownership of land by individuals." He concluded by stating, "The existence of economic incentives and personal liberties deter-

Opening ceremonies in 1953 featured the largest American flag ever suspended in the air. Measuring 60 x 80 feet, it fluttered above the heads of four high school bands. Courtesy Oil and Gas Journal.

Charlton H. Lyons, president of the Independent Petroleum Association of America, helped open the IPE's 30th anniversary show with his keynote address. Courtesy Oil and Gas Journal.

mine the efficient development of all natural resources. When men are free to think, to argue, to dissent, to dream, there is no limit to their ability to produce."[58]

Governor Murray welcomed the foreign visitors, particularly those from Latin America. Fluent in Spanish, Murray addressed these delegates in their native language to their evident delight. Mayor George Street of Enid followed the governor and presented a plaque to Skelly on behalf of the oil producers of southern Oklahoma; they were honoring Skelly for his 40 years of service to the oil industry. As these ceremonies concluded, there was a thunderous roar as engines were moved to full throttle marking the opening of the 30th anniversary of the IPE.[59]

Scattered across the grounds were rigs of various heights and types, some of them capable of drilling to great depths, others highly mobile, slimhole rigs—even a variety of spudders. The mast-dotted 26 acre site had equipment and displays valued at $100 million; there were 1484 exhibitors representing every phase of the oil industry.[60] Non-industry people were impressed by the height, weight, and bulk of the giant tools and engines, but industry experts were impressed by how far the technology of their business had come in the five years since the show of 1948.

They also were impressed with the giant strides made by foreign manufacturers, represented at the IPE for the first time in a major way. Prior to the 1950s most foreign oil firms had purchased American equipment, but challenging this tradition were such firms as Eisenwvrkwuefel of Germany, *Societe Nationale de Material Recherche L'Exploration* of France, and the Council of British Manufacturers of Petroleum Equipment. Moreover, the foreign firms displaying at the IPE in 1953 gave every indication that they were going to become a permanent part of the American scene.[61]

An outstanding feature of the 1953 exhibits was the striking demonstration in several ways of how the petroleum industry had adapted developments in other areas to its own uses. Electronics had invaded nearly every operation and process performed by oil men; pipeliners were using microwave communications systems; field operations personnel were using mobile radios; surveyors were using electronic direction finders; computers were being used to solve mathematical problems and for analyses; and electronic devices for measuring and controlling operations were being used from airplanes and ships. Moreover, atomic power had been added for use in well-logging equipment, for tracing pipeline operations, and for analyzing refining processes.[62]

Advances in metallurgy were much in evidence at 1953's exhibits, for the oil industry depended heavily on a constant supply of lighter, stronger materials. Every new alloy was examined with intense interest to determine how well it would stand up under stress and corrosion and how long it would last. Newer metals, such as aluminum and magnesium, also were to be found in many exhibits with glowing claims made for their special characteristics.[63] Alongside these exhibits of metals were plastic products made possible by the petrochemical industry. This versatile substance was becoming a staple of use in the oil industry, for it could be seen in well-bottom equipment, coatings, pipes, and corrosion-resistant tanks.[64]

Several exhibits featured new developments in the science of hydraulics. These made possible jet bits that directed drilling mud in such a way as to wash cuttings away from the bit teeth and increase drilling speed. Better knowledge of hydraulics also could be seen in improved pumps, valves, conduits for drilling mud, refinery catalysts, and high-pressure gasses.[65]

Another industrial innovation appropriated by petroleum was automation. Oil men were intrigued by the many new instruments and devices offering the twin blessings of reduced human effort and reduced human error. Items on display in 1953 included automatic computers, calculators, electronic analyzers, and remote-control devices. "The old-time driller," wrote Henry Ralph in the *Oil and Gas Journal*,

> who went by the sound and the feel of the drilling cable, and the early day refiner, who operated his stills by the seat of his pants, would be flabergasted at the array of instruments of all kinds offered for every phase of petroleum operations. Instruments, gauges, meters, recorders, analyzers and automatic adjusters are the order of the day in this 1953 equipment exhibit, exemplifying the observation that this is the age of instrumentation."[66]

However, smaller, delicate, more complex equipment did not mean the end of the oil field behemoths with which oil men felt comfortable. There still were giant machines crawling and rumbling across the IPE grounds in 1953. Great earthmovers towered over the exhibit scene, and there were giant cranes, bulldozers, ditchers, and tractors aplenty.[67]

By the time oil men had toured the exhibit booths and circled the giant machines parked outside, they had armloads of pamphlets, brochures, and other promotional literature explaining what they had seen. For them this material was "must reading," as was the continuing stream of new information coming from trade presses each month. One exhibitor, the *Oil and Gas Journal*, tried to help by offering free "Improve Your Reading" tests to help busy executives get through the daily "engineering reports, business letters, interoffice memoranda, [and] business magazines." As the *Journal* noted, "It has somewhat suddenly dawned on alert management that a considerable part of the higher-level personnel efficiency rests upon how fast they can read, and

U.S. Steel in 1953 displayed a model of a new barge developed for over water drilling. Courtesy Oil and Gas Journal.

Professor Lois DeFigh of the University of Tulsa's Reading Clinic demonstrating reading improvement techniques to petroleum engineering students at the 1953 IPE. Courtesy Oil and Gas Journal.

the proven fact that greater speed in reading can fairly readily be taught."[68] Many oil men came away from these tests convinced that faster reading was possible and necessary, after which they moved on to the Hall of Science—where they could fill yet another shopping bag with brochures and information booklets.

In the Hall of Science oil men found the emphasis on recent scientific and technical advances. The criterion for selection for inclusion in the Hall of Science was the degree to which a prospective display reflected advances made in the last five years. The Hall of Science committee, headed by G.H. Westby and including Kent W. Kimball, Dr. B.B. Weatherby, A.E. Ballin, and Dr. Gustav Egloff, believed that developments were happening so fast that only a five-year span could receive comprehensive coverage. Exhibits in the Hall were the product of a worldwide search by the committee "for exhibits, displays, processes, equipment, and scientific presentations that [would] dramatize and demonstrate all phases of the petroleum industry's operations and progress." Companies vied for inclusion in the Hall of Science exhibit, for this was considered a signal honor. Space was allotted without charge, and all exhibits were fully insured.[69]

In 1953 for the first time the Hall of Science was partitioned to direct visitors through a logical sequence of the history of oil from its discovery. An exhibit on geology was first in the progression and featured a diorama that showed the surface expression of the rock structure and a cross-section in the foreground; under this cross section was a lighted geologic timetable showing the development of life through the geologic ages. Another panel illustrated on-going surface processes such as erosion and deposition and the accumulation of oil in traps.[70] The next exhibit highlighted the uses of geophysics in the oil industry, while animated models showed the principles of seismic, gravity meter, and magnetometer surveying.[71]

Oil field development and production came next among the partitioned exhibits. One of the most popular exhibits in the Hall was in this section, a working scale model of a drilling rig actually "making hole." Stanolind Oil and Gas Company contributed working models of two new production methods, one showing a new gas-lift method of producing oil and the other showing the first fully automatic paraffin scraper.[72] Fourth among the exhibits was the transportation division's demonstration of progress in pipelining during the past five years. Also shown in this exhibit was microwave communication with a working installation in the Hall; six circuits were connected between installations at opposite ends of the Hall, three of them connected to telephones, one to a teletype, and the others working remote-controlled valves.[73]

Refining was represented by an operating model of a refinery which took crude from a flowing well and separated it into a variety of petroleum products, all in a transparent tower so visitors could watch each step as it was performed. Another display in this division showed the various kinds of crude with which refiners contended, and there was a demonstration of an electron microscope.[74] At the last station in the Hall, visitors saw an exhibit displaying the various petrochemicals being made in 1953. A giant drawing of the "Growing Tree of Petrochemicals" showed the roots in oil and gas fields and its branches labeled "detergents, additives, sulfur and sulfur derivatives, refrigerants, medicines, solvents, antifreeze, dyes, agricultural chemicals, plastics, carbon black, synthetic rubber, synthetic fiber, and synthetic resins." The top branches were labeled "countless petrochemicals still to come."[75] In this same exhibit were displays showing the many consumer products made possible by petrochemistry.[76]

After touring the Hall of Science, visitors could go to a nearby theater and select from more than 50 films available for viewing on subjects such as exploration, production, transportation, refining, research, science, and education. This theater was open daily from 11:00 a.m. to 9:30 p.m., and the day's program was posted near the entrance so visitors could come and go as they wished.[77]

Many other forms of entertainment were available. The IPE always generated many things for visitors to see and do including symphony concerts, baseball games, tours, and civic club luncheons. Three concerts by the Tulsa Philharmonic Orchestra took place on Drake Drive in front of the Texas Building. For wives of visiting oil men there was a tour of Woolaroc Museum and a luncheon in Bartlesville sponsored by the local Desk & Derrick Club and Phillips Petroleum Company; a style show and tea at the Tulsa Petroleum Club; or a visit to the Will Rogers Memorial at Claremore.[78]

For foreign visitors there was an even greater variety of entertainment. First came an informal party for them on May 20 at Southern Hills Country Club hosted by W.H. Helmerich, president of Helmerich & Payne Drilling Company. Others hosted a dinner for foreign dignitaries at the Tulsa Club, while Governor Johnston Murray and his wife greeted foreign visitors at an open house at the Tri-State Building penthouse. And the NOMADS hosted a cocktail party and buffet dinner at the Cimarron Ballroom.[79] There were many foreign guests in 1953, for as Manager W.B. Way had said a year before of the effort to attract visitors from other countries, "We've made more contracts further in advance of this show than for any of the previous expositions."[80] He added, "Down through the years, foreign governments have been compelled to give more and more recognition to this show because of the tremendous opportunity it alone provides their representatives to get acquainted with the petroleum industry—its techniques, equipment and its outstanding contributors."[81]

Dr. Oscar B. Irizarry had gone to Washington, D.C., to visit foreign diplomats, ambassadors, and ministers, then had made a 92-day tour of South America. Accompanying him on many of these visits was Governor Johnston Murray, and together they presented more than 600 invitations to the IPE. Others making foreign contacts on behalf of the IPE included H.R. Keplinger, C.J. Hochenauer, president of Texoma Supply Company, and Carl White, Jr., president of Franks Manufacturing Company.[82]

The NOMADS, as usual, met deplaning foreign visitors at Tulsa Municipal Airport and coordinated their activities until they were ready to return home. Oppie Dimmick, president of Tulsa NOMADS in 1953, had his committees organized by the time the first reservation was made that year by three men from Hamburg, Germany. These commitees remained at "full throttle" until all details had been attended for registering, housing, guiding, receiving, and entertaining visitors from foreign nations. Among the countries represented in 1953 were Canada, England, Venezuela, Brazil, France, Germany, Italy, Mexico, Scotland, and the Netherlands, all of which entered exhibits in the show.[83]

Foreign exhibitors at the IPE—numbering 36 in 1953—were a recent phenomenon, but not the only one in 1953. This was the first year the title "Woman of Achievement was awarded."[84] Competition for this title was directed by the Desk & Derrick Clubs of North America "in recognition of the mounting contributions of women in the many phases of petroleum industry operations."[85] Desk & Derrick women in 46 chapters, located in 17 states and Canada, participated in this selection. Desk & Derrick clubs had originated two years previously in New Orleans, and in 1952 the Tulsa chapter proposed an award for the "Woman of Achievement." Letha Dillon of Deep Rock Oil Corporation appointed a committee headed by June Gregory of Sneed Oil Company to honor the first Woman of Achievement. Mrs. Gregory's committee set a deadline of March 31, 1953, for nominations, and by mid-April Ernestine Adams, an oil magazine editor from Dallas, had been selected as the first recipient of this honor. She received an expense-paid trip to the IPE and was honored at an elegant banquet.[86]

Others honored that year included: Harry J. Crawford of Emlenton, Pennsylvania, as "Pioneer of Pioneers"; W.L. Connelly of Tulsa as the "Grand Old Man of Production"; Edwin B. Reeser of Tulsa as the "Grand Old Man of Refining"; Dr. George G. Oberfell of Bartlesville, "Grand Old Man of Natural Gasoline"; Dr. Goddrey L. Cabot, "Grand Old Man of Natural Gas"; and Wallace D.

Oklahoma Governor Johnston Murray shows W.G. Skelly (second from right) one of the invitations to the IPE he extended to Latin American officials. Looking on, L to R, are: E.B. Reeser, C.H. Pape, O.B. Irizarry, and Alf G. Heggem. Courtesy Oil and Gas Journal.

Wilson, "Grand Old Man of Supplies and Equipment."[87]

In 1953, the 30th anniversary of the first show, the IPE proved as vital as at any previous show. It was still developing, adapting, changing—and setting records. The "Woman of Achievement" award started an important new tradition; the new 10-day format had been successful; and 450,000 visitors, 28,000 of them oil men from 38 states, had pushed through the turnstiles of the exposition grounds.[88]

1959

In Tulsa on the opening day of the IPE in 1959 was Ruth Sheldon Knowles, whose book, *The Greatest Gamblers*, was being released to mark the 100th anniversary of the oil industry. She dedicated her book "with gratitude and admiration" for the "courage, venturesomeness, faith, persistence, and optimism" of "all the unsuccessful explorers who have drilled America's 300,000 dry holes and whose failures have guided others to the discovery of an abundance of oil."[89]

The traits Mrs. Knowles found in these men certainly had been present in the man who had led the IPE for 30 years, W.G. Skelly. His death in 1957 presented the IPE with a crisis of leadership. After a frantic search, the executive committee selected William K. Warren, who was Skelly's equal in determination, energy, and leadership. At the time of his selection, Warren was chairman of the board and chief executive officer of Warren Petroleum Corporation and a director of Gulf Oil Corporation.[90] He also was a respected civic leader and philanthropist who had served on the board of directors of the IPE for more than 20 years and had been a member of the executive committee since 1952.[91]

Born in Nashville, Tennessee, Warren had moved to Tulsa in 1916 where, after working for Gulf, Margay, and the McMan oil companies, he founded Warren Petroleum Company on March 15, 1922. This and other firms he started were merged into Warren Petroleum Corporation in 1937. He was a member of the American Petroleum Institute, the National Petroleum Council, and the Mid Continent Oil and Gas Association, as well as a director of the Tulsa Chamber of Com-

W.K. Warren became president of the IPE in 1957 and served for 11 years. Courtesy Oklahoma Heritage Association.

merce, director of the First National Bank and Trust Company of Tulsa, a trustee of the University of Tulsa, a member of the College of Commerce Advisory Council at the University of Notre Dame, chairman of the board of St. John Vianney Training School, a trustee of St. Francis Hospital, Inc. (founded by the William K. Warren Foundation); and a director of the Missouri-Kansas and Texas Railway.

Also selected as an IPE officer in the summer of 1957 was J.L. Shakely, who was named treasurer to succeed Clyde Pape.[92] Warren and Shakely joined other officials of the IPE to begin preparing for what was expected to be the grandest IPE ever held, the 100th anniversary show of the beginnings of the oil industry. Other members of the executive committee included E.F. Bullard, D.D. Bovaird, A.W. McKinney, Walter H. Helmerich (the newest member), William B. Way, and H.R. Powers. Also assisting with this show were George Roberts, Jr., chairman of the scientific and technical committee; Dr. A.L. Levorsen, chairman of Exploration Day at the show; A.W. Thompson, chairman of Drilling Day; Jake L. Hamon, chairman of Production Day; Reid Brazell, chairman of Refining Day; J.L. Irwin, chairman of Pipeline Day; Charles E. Webber, chairman of Natural Gasoline Day; John F. Merriam, chairman of Natural Gas Day; William J. Sherry, chairman of the Old Timers committee; D.A. Bridenthal, IPE assistant manager; A.F. Keating, chairman of the IPE housing committee; S.J. Raphel, manager of the IPE housing bureau; Leslie Brookes, IPE publicity director; and H.B. Baber, Jr., IPE assistant publicity director.[93]

In April of 1959 President Warren and C.H. "Kep" Keplinger, engineer and consultant, met with representatives of the Natural Gas Association to discuss the upcoming IPE. Each show, they explained, had been larger than its predecessor, and from all indications the 1959 show would continue the pattern, all available exhibit space having been sold since the end of January. In his presentation Keplinger referred to maps which he had prepared and placed at tables to show what was happening around the world in oil.[94]

In 1918, he said, Tulsa was truly the Oil Capitol of the World and, of the one million barrels of oil produced in the world, more than 40 percent of it was coming from around Tulsa. In 1959, oil production had doubled around Tulsa, but the percentage figures had changed. In 1958 the entire world produced 17,700,000 barrels of oil per day, and there were 275,000,000 barrels in worldwide reserves, 60 percent of which was American controlled. Tulsa, even though its percentage of this production had dropped sharply, continued to lead in oil exploration, research techniques, drilling and transportation. This leadership, according to Keplinger, was largely attributable to such institutions as the University of Tulsa's Petroleum Engineering School and the *Oil and Gas Journal*.[95]

Another civic leader, Burch Mayo, spoke of Tulsa's debt to the oil industry and of the gratitude Tulsans should feel on the 100th anniversary of the oil industry to be the "First City of Oil":

> It's a good time for Tulsans to take inventory of their city and see how much each of them owes to the oil industry....Right now Tulsa has

more than 125,000 persons employed in non-agricultural jobs....If the oil strikes of the early 1900s and the subsequent headquartering here of hundreds of oil companies had not spurred Tulsa's growth at such a fantastic pace, Tulsa probably would be a minor area trade center with 5000 or so jobs. In other words, if this is valid reasoning, approximately 24 of every 25 employed workers in Tulsa owe their jobs— directly or indirectly—to oil.[96]

It probably was not necessary to remind Tulsans of their debt to oil inasmuch as their city always had displayed unstinting pride in its petroleum heritage. Tulsans repeatedly had risen to the challenge of housing, feeding, and entertaining the thousands of oil people who flocked there when the IPE opened its gates. Some 30,000 came daily in 1959, and more than 2000 Tulsa families would open their homes to provide lodging for them. The IPE housing committee, headed by Anthony F. Keating, coordinated reservations for the show to keep confusion to a minimum and to assure that every visitor had accommodations, an effort aided by the IPE's 10-day format. Under this plan, companies were urged to stagger attendance of their personnel to allow more people to see the IPE, to keep back-home operations going, and to ease the housing problem in Tulsa.[97]

As the housing and exhibits committees were attempting to accommodate the crowds, other groups were working to make the crowds even larger. The IPE press bureau had been set up on the exposition grounds in an air conditioned room complete with the latest facilities. Through the use of a teletype, the wire service, and the wire photo service, reporters for the first time could write and file their stories from one location on the IPE grounds. This press room was staffed with five IPE employees under the direction of Leslie Brooks, publicity director, and C.W. Nicholson, assistant publicity director. IPE officials had intended to have the press room ready but not open until May 14, but the interest in the 100th anniversary IPE was so great that the room had to be opened early so reporters could begin filing background stories and collecting the materials they would use in forthcoming issues.

Publicity for the show already had been underway since 1957, according to Brooks.[98] The campaign had been conducted with releases to 14 foreign and 54 domestic trade publications, 521 newspapers, 184 radio stations, 63 television stations, and 68 house organs of oil companies and associations. Brooks reported that his staff had distributed 600,000 mail stuffers, 200,000 brochures, 27,500 bumper stickers, and 75,000 windshield stickers. All this meant that Tulsa had "an opportunity of a lifetime in 1959 to publicize itself as the Oil Capitol of the World...."[99] This prediction was made by W.L. Connelly, retired Sinclair Oil Company official, who added, "By all measurements Tulsa will have the largest gathering of oil men at one time of any city in 1959."[100]

The usual parade of exhibitors from downtown to the showgrounds marked the beginning of the IPE in 1959. At the IPE 1497 exhibitors were waiting to display half a billion dollars worth of equipment in some 1000 booths and 26 acres of outside exhibit space. The official directory that year listed names ranging from Aaron Machinery Company to Yuba Heat Transfer Division, and each was on hand with its newest and latest wares. Edwin Drake would have been astounded at what he had started on August 27, 1859. When he died at Bethlehem, Pennsylvania in 1880, an object of pity dependent on public charity, the oil industry for him had been a bitter disappointment and the $2000 he had spent drilling that first well a total loss. On May 14, 1959, suppliers of the worldwide oil industry had multiplied Drake's investment 250,000 times in preparing for a single display of oil field tools, supplies, and equipment.[101]

On Drake Drive, in a "colorful, impressive ceremony, and with the representatives of numerous foreign governments and companies, the greatest oil show of them all was formally opened."[102] From the speaker's stand just inside the gates, spectators and guests heard greetings from President Dwight D. Eisenhower; a message of oil industry well-being from General Ernest O. Thompson of Texas; greetings from Ernesto Peters, a government oil representative from Argentina; and welcomes from IPE President W.K. Warren, Tulsa Mayor James L. Maxwell, and Oklahoma Governor J. Howard Edmondson.[103]

President Eisenhower's message was delivered by Wendell B. Barnes, administrator of the Small Business Administration and a former Tulsan, who declared, "Today my happiness, my pride know no bounds, for today Tulsa is truly a world city, host to the greatest international trade fair in

Office of the IPE in 1959. Courtesy Petroleum Engineer.

existence on the face of the globe."[104] Ernesto Peters, general manager of production and drilling by the Argentine government, spoke more expansively: "This tremendous and unique display of oil equipment is symbolic of the industrial might of this great republic....Furthermore, it is a tangible demonstration of the part that business can play in promoting international understanding."[105] General Thompson spoke about the "brilliant prospects" of the oil industry but cautioned that conservation still had to be the industry's watchword and that refiners had to guard against contribution to dangerous over-supply. Governor Edmondson expressed a general greeting, while Mayor Maxwell urged visitors to consider Tulsa "the second best city in the world, second only to your own hometown."[106]

President Warren then officially opened the show, dedicating the IPE commemorating the 100th anniversary of the oil industry to his predecessor, W.G. Skelly, saying, "From the day that Bill Skelly became president of the IPE in 1925 until his death, his was the hand that steered this exposition show by show until it became recognized as the greatest single industry exhibition in the world. Therefore I think it is fitting that in memory of Bill Skelly I now declare this International Petroleum Exposition officially open."[107] As the crowd dispersed into the streets and walkways of the grounds, "American flags and gay colored pennants fluttered in the breeze to a background of blatant diesel exhausts and other equipment."[108]

Amid the noise a 7500-pound gold-plated "derrick man" appeared to climb the 136-foot rig erected by Mid Continent Supply Company to get a better view of the expansive array of equipment and machinery below him—at a time when there were fearful predictions from some that the oil industry was in a state of decline as the world's premier energy source because of the promise of nuclear power. However, Frank M. Porter, president of the American Petroleum Institute, spoke for most oil men when he stated that oil was not "nearing the end of the road. The electric light which supplanted kerosene did not extinguish the oil industry,...but merely preceded the development of automobiles, airplanes, natural gas pipelines and petrochemicals."[110] These developments, he pointed out, expanded the oil industry to "undreamed of" dimensions, and he predicted that nuclear energy would do likewise. His theme was that predicitons of the demise of the oil industry would always be premature, and the variety of complex and innovative exhibits on the exposition grounds bore him out. Improved efficiency, economy, and mobility were to be found at nearly every exhibit.

Some of these innovations were deceptively simple yet momentous in their implications. Within

Workmen at the refurbished Oklahoma Building in 1959. Courtesy Oil and Gas Journal.

the past year world oil production had been increased significantly by the simple expedient of injecting into wells "detergents similar to those used in cleaning dirty clothes."[111] Thereby thousands of wells had been restored to production, adding millions of barrels to potential reserves. The secret was that detergents reduced the surface tension between oil and water underground. R.M. Lasater, a chemist for Halliburton Oil Well Cementing Company, exhibitor of the detergent, described it as "an oil soluble anionic surface active agent"[112] which had proved successful for both old and new wells. Testing of the substance on display had been completed just before the 1959 IPE opened.[113]

Efficiency, along with versatility and portability, could be seen in Continental Emsco's gas turbine engine. It was bidding to replace the standard reciprocating engine used on most rigs. At the IPE the company had a portable rig with a 500-horsepower Solar Aircraft Company turbine displayed. According to O.M. Sievert, assistant manager of turbine and control sales for Solar, the chief advantages of the turbine were low weight

William G. Marmack, left, moved from Pittsburgh to Tulsa with his family, rented an apartment, and spent six and one half months to setting up National Supply Company's exhibit. A change in the blueprint before the show brought this huddle. Courtesy Oil and Gas Journal.

per horsepower, fuel versatility, and lack of vibration. Its drawback was its high cost due to limited production.[114]

Parkersburg Rig and Reel Company of Coffeyville, Kansas, was touting a worm-screw oil well pump featuring light weight and high efficiency. Called the "X-D Pumping Unit," it had a direct-drive mechanism with a motor on top to create a wormscrew action which developed a stroke up to 40 feet and would pause at the bottom of its stroke for up to 30 seconds to permit the pumping barrel to fill. R.H. Hill, one of its developers, claimed that this pump needed only some 60 percent of the power of the hydraulic-lift pump. The unit on display weighted only 2800 pounds compared to the 28,000 to 30,000 pounds for rocker-arm pumps.[115]

Lighter weight also was a feature at Reynolds Metals Company's display of 4.5-inch aluminum drill pipe. At the time of the 1959 IPE, four major oil companies were testing aluminum drill pipe, and the early results had surprised even the manufacturer. Steel pipe had reached below 25,000 feet but there were serious weight problems. Aluminum required more metal but had only half the weight of steel, and early results indicated that the lighter drill pipe could increase depth capacity of drilling rigs by as much as 60 percent. However, it did cost more.[116]

Mobility was featured in an exhibit jointly sponsored by Seismic Service Corporation, Republic Aviation Corporation, and Allied Helicopter Service. Each day a portable seismic rig was flown by helicopter from Seismic Service's headquarters at 1600 East 41st Street to the IPE grounds where a two-man crew set it up and began drilling a short hole. The helicopter used was the Alouette, the world's first production gas-turbine aircraft.[117] This event was a favorite with spectators. A different exhibit, equally popular, was sponsored by Dowell, Incorporated: a Nike Ajax surface-to-air missile. This 32-foot, 2000-pound rocket had been brought to Tulsa from Fort Bliss, Texas, on two large trailers and was shown as part of Armed Services Week activities. Lieutenant Colonel E.H. Kyle, Army Deputy Information Officer, was a guest at the Dowell exhibit and provided information about the missile to intrigued spectators.[118]

From automobile manufacturers came exciting exhibits. Ford Motor Company, in a display covering more than 10,000 square feet, had 10 different types of trucks used in the oil industry, engines and power units, and industrial tractors. The Berliet Company of France had brought the world's largest truck, a vehicle designed for use on the Sahara Desert; it was named the "Tulsa" in honor of the IPE and the 100th anniversary of the oil industry.[119]

Less glamorous was the exhibit of the Hughes Tool Company. On display was the drill bit invented by Howard Hughes, Sr., and used at Goose Creek, Texas, in 1909, along with the RG-1 series bit used in 1958 to drill to 25,340 feet, then the world's record.[120] The Garnder Denver Company had on display a new all-purpose pump designed for use on drilling rigs using mud servicing, cementing, fracturing, workover operations, or whatever. Ramsey Winch Company was displaying a wide array of its products in 30 exhibits, from the "Baby Ramsey" with its 1000-pound capacity to the "Giant Ramsey" with its pulling power of 100,000 pounds.[121] Perhaps the most nostalgic exhibit in 1959 was the small booth of Myron M. Kinley, the legendary fighter of oil well fires. For years presidents of oil companies had carried his telephone number, and when his phone rang it

The IPE's giant flag waves over an afternoon session of the 1959 show. Also visible are the renovated Texas Building and individual exhibits by General Motors, Bethlehem Supply Company, and Mid-Continent Supply's huge derrick. There were 16 derricks on the grounds in 1959, the last year at which the big derricks were shown. Courtesy Leslie Brooks.

meant a trip to an oil patch somewhere in the world where there was trouble. Although by 1959 he was slightly deaf and somewhat lame, he was treated with the respect and deference usually reserved for nobility by the 30,000 oil men who came to the IPE in 1959.[122]

Awaiting visitors in the fabulous Hall of Science were the "latest oil industry scientific advances backstopped by notable milestones of the past."[123] One of these was the perennial exhibit of Colonel Drake's drilling tools. "Latest advances" included a model offshore drilling rig, a working model of a caisson for repairing underwater pipelines, new instruments for exploration, an earthquake recording drum, and sound records of an earthquake and an atomic bomb explosion. In a before-and-after exhibit of seismographic equipment, Marland Oil Company's early-day reflection equipment was contrasted with the newest transistorized equipment.[124]

The production division of the Hall of Science showed the change in production procedures over a five-year period, weaving a complete story showing the drilling of a well and reservoir engineering. Another display showed models of spring pole drilling, an early cable tool rig, a modern cable tool rig, and the newest rotary drilling rig.[125] In displays reflecting the testing and completion phases of oil field work, there were new tools for coring and drillstem testing, wire line formation testing equipment and animated models showing the principles of formation fracturing, and models showing the latest developments in fracturing procedures. Pipelines were represented by a working model of an automated pipeline system complete with a dispatching board from which the system could be controlled remotely.[126] In the area of the Hall of Science devoted to refining, visitors gathered around an 8 x 16-foot model of a 230,000-barrel-a-day refinery which showed how refiners were building plants which did a better job at lower cost.[127]

A series of movies were running continuously between 10:00 a.m. and 10:00 p.m. in the small theater adjacent to the Hall of Science giving visitors a comprehensive view of the oil industry. This theater could accommodate only 120 people, and it was filled to capacity at almost every showing. This theater was a natural last stop for visitors whose interest had been piqued by exhibits in the Hall of Science[128]

Some 2000 of these visitors were from foreign lands. They were guests of the NOMADS organization which gave each foreign delegate a large pliofilm badge showing his name, country, and company. By using these badges for foreigners, NOMADS hoped to provide them a convenient

L to R: Police Chief Joe McQuire, Police Commissioner Robert Mawhinney, and Captain Louis J. Skinner study "mugs" of doubtful citizens expected in Tulsa for the IPE in 1959.

Roberto Valverde, chancellor of the National School of Engineering at Lima, Peru, arriving in 1959. Greeting him was Oscar B. Irizarry, NOMADS president, along with (L to R) Herbert C. Fries, John Pearce, and Ben F. Kelly of the NOMADS greeting committee. Courtesy Oil and Gas Journal.

means of identifying themselves and of getting assistance should they need it.[129]

Another group with a major responsiblity was the Tulsa Chapter of the Desk & Derrick Clubs of North America. It served not only as housing counselor for IPE delegates and sales persons of IPE novelties on the grounds, but also was host for Desk & Derrick members from across the country who had come to see their "Oil Woman of the Year" honored. May 16 was designated "Desk & Derrick Day" at the exposition, and honored as "Oil Woman of the Year" in 1959 was Elizabeth Aldrich Bridgeman, a fuels technologist in the research and development department of Phillips

Dr. O.B. Irizarry extending an invitation to the 1959 IPE to Brazilian ambassador Ernani do Amaral Peixoto in Washington, D.C. Courtesy Oil and Gas Journal.

Petroleum Company of Bartlesville. She was honored at a banquet in the Crystal Ballroom of the Mayo Hotel where she was presented a diamond wristwatch. Other women receiving "honorable mention" certificates were Irma Cline, owner of Petroleum Business Services, Inc.; Bernice W. Cleers, oil editor of the San Angelo (Texas) *Stand-ard-Times*; and Margaret Applegate Kitchen, senior geologist with Sun Oil Company of Mount Pleasant, Michigan.[130]

Old Timers honored in 1959 included William L. Connelly of Tulsa, one of the founders of Sinclair Oil and Refining Company; he was named "Pioneer of Pioneers" in recognition of his distinguished 64-year career. The other seven "Grand Old Men" included Edmond C. Breene, 81, of Oil City, Pennsylvania, in production; William A. Cassidy, 93, of Bayonne, New Jersey, in refining; Edward I. Hanlon, 78, of Tulsa in natural gasoline; Frank J. Hinderliter, 84, of Tulsa in supplies and equipment; Wiley B. Hissom, 66, of Tulsa in drilling; Frank J. Lerch, 71, of New York, in natural gas; and S. Miller Williams, 72, of Robinson, Illinois, in pipeline and transportation.[132]

These pioneers were honored on May 17 at a ceremony held beside the replica of the Drake Well. Speaker at the event was Dr. Paul Giddens, president of Hamline University, and all members of the Old Timers organization were given gold lapel pins to signify their special status among the more than 30,000 oil men who attended the 14th IPE along with 2042 foreign oil men (three times the previous record) and 547,208 assorted visitors in attendance. These totals far exceeded the record set in 1953 when 395,352 had twirled the turnstiles of the admission gates to wander through the oil industry's fantasy land.[133]

CHAPTER FOUR

FINAL YEARS
(1966-1979)

In 1966 the IPE moved to a new 10.5-acre building and hired a professional manager from Chicago to put together a streamlined package for exhibitors in order to reduce the cost of participating in the show. Exhibitors increasingly were demanding a better ratio between the cost of exhibiting and their return from sales. They wanted shorter shows, more exposure to genuine buyers, and less interference from the sight-seeing public.

Although pressure from exhibitors was increasing with each show, most Tulsans and several IPE officials strongly opposed barring the general public from the technical exhibits.[1] The IPE finally responded to the demands of its constituents by designating certain days of the show for industry people and other days for the public, and the length was gradually cut to just four days. Finally, the IPE came under pressure from competing shows and began to experience losses that had to be covered by its reserve fund. The alternatives were clear. The IPE could continue to compromise its basic format as a non-profit, industry owned exposition and compete with other oil shows (spawned by the success of the IPE), or it could step aside and enter history as the one and only "World's Fair of the Oil Industry." The ultimate decision came on April 7, 1979.

1966

In August of 1960 the Petroleum Equipment Suppliers Association voted that there was an "imperative need" to reduce expenses in the face of a business slump combined with a proliferation of oil shows around the country. Thus it resolved: "that the Board of Directors of the Petroleum Equipment Suppliers Association recommends to the Board of Directors of the International Petroleum Exposition and to the Boards of Directors of all other oil shows that the Tulsa oil show and all similar oil shows be discontinued."[2] This resolution went to officials of the IPE, the Louisiana Gulf Coast Exposition at Lafayette, Louisiana, and the Permian Basin Oil Show at Odessa,

Martin Dwyer, who succeeded William B. Way as IPE manager. Courtesy Oil and Gas Journal.

Texas. Since 1953 the domestic oil industry had been plagued by a market glut, and both big and little suppliers had been "scrambling for business and carefully watching for places to save money."[3]

IPE officials responded that "feast-or-famine conditions have appeared periodically in the past" and that "rapid technological development in oil equipment in the six-year period from 1959 to 1965 [he date first projected for the 15th IPE] will probably produce enough new equipment that exhibitors will want to show it."[4] IPE President W.K. Warren announced that, although he fully understood the reluctance of PESA members, plans for the 15th IPE would go forward.[5]

Warren was fully aware that if Tulsa did not host the oil show other cities would, for several rivals were clamoring to hold the big international show. On October 29, 1963, a group of business leaders— bankers, oil men, state fair officials, and officers of the Chamber of Commerce—met at the Tulsa Club to discuss ways to keep the IPE in their city. Their conclusion was that the "old and infirm" exhibition buildings no longer would support "the new character of the show" as a no-frills, strictly business trade show.[6] Their solution was to propose a $3.5 million bond issue to be held on December 3 with the proceeds used to construct a 10.5-acre exposition building on the Tulsa State Fairgrounds.

Martin C. Dwyer, president of the Chicago trade show management firm hired to oversee operation of the IPE, explained to Tulsans that this new building was needed because exhibitors no longer could afford to build their own structures (as they had in 1953 and 1959) and then let these stand idle during the five- or six-year interval between shows. He also explained that new developments in instrumentation, automation, and processes would change the character of the IPE from an exterior to an interior show.[7]

Another break with tradition was the lease signed between the Tulsa State Fair and the International Petroleum Exposition Corporation linking housing for the IPE and the fair under one management. G.C. Parker, president of the Tulsa Exposition and Fair Corporation, announced the new agreement and explained the terms of the lease:

> The lease assures keeping the IPE in Tulsa and provides for a progressive expansion and improvement program for both the IPE and the Fair.... The lease states the IPE may hold one Exposition during the month of May in each five-year period. If an oil show has not been held during a ten-year period, the grounds revert to the Fair. The Fair Corporation will maintain the grounds and the two corporations are empowered to construct buildings on a joint basis if this is feasible. The Fair may also lease or use any building belonging to the IPE either during the annual Fair or for rental or storage during the off-season. The IPE may use any part of the adjoining fairgrounds for its expositions. Rental will be $2500 each year the IPE stages a show plus ten percent of the gross income from exhibit space. No off-season maintenance charge will be made to firms or individuals owning buildings on the grounds providing the buildings are made available to the Fair for use.[8]

The Fair and the IPE were tied into one "set-up," Parker explained, thereby saving the IPE up to $30,000 a year in maintenance costs. "In the past," he said, "we have been two families in the same house.... Under the new agreement things will be better for us both."[9] In an aside he noted that long-time IPE manager W.B. Way, who had lived for years in a house on the IPE grounds, had moved.

While these negotiations were underway, IPE supporters had mobilized their forces to push passage of the bond issue. Board members of both the Fair Corporation and the IPE agreed to raise the $13,000 needed to finance the election. W.K. Warren reminded Tulsans that the IPE would pull some $40 million in new money into the community and that without the proposed facility these monies would be lost after the 1966 show. Other civic leaders met with clubs, forums, and citizens' groups to lobby for the bond issue as a means of helping Tulsa's industrial community.

In Chicago at a meeting of oil writers, Tulsans were quizzed closely about the prospects of success for the bond issue, and these Tulsans returned home convinced that if the issue failed other cities, principally Dallas and Houston, were waiting to become host to the IPE. When at last the ballots were counted, 25,949 had voted for it, 11,876 against it.[10] Tulsa's hold on the IPE was assured—at least for the present. On December 4, 1963, Tulsa Fair Corporation President Parker announced that detailed plans for the new building would be published in approximately six months and that ground-clearing at the site would begin in mid-1964. It would be on space previously occupied by IPE buildings, whose salvage would pay the cost of clearing the area.[11]

On April 3, 1966, an estimated 5000 people were on hand for the dedication of a "new Tulsa industry," the "gargantuan" IPE building at the fairgrounds. Governor Henry Bellmon, speaking from a platform "loaded with local political and oil dignitaries referred to the 'super duper' structure as a new industry and said Oklahomans ought to brag more...maybe not as much as Texans, but more."[12] W.K. Warren thanked Tulsans for supporting the bond issue and said the new building would help keep Tulsa in the forefront of the oil industry.[13]

The giant building, when completed, was four blocks long, encompassed 10.5 acres, and incorporated a cable suspension roof system that eliminated interior pillars. Elaborate water, gas, and

The Tulsa Exposition Center, built for the IPE by the citizens of Tulsa. The 10.5-acre building still stands and is used for many purposes. Courtesy Leslie Brooks.

Oklahoma Governor Henry Bellmon congratulates IPE officers (L to R) Helmerich, Shakely, and Warren on completion of the Tulsa Exposition Center.

electrical systems in it could support almost any exhibit, and moveable exterior wall panels enabled large pieces of equipment to be brought into the structure from almost any point. Railway sidings paralleled all entrances. This mammoth structure, the largest of its type in the nation, replaced the old Oklahoma, Texas, Kansas, California, and Technical buildings owned by the IPE along with those owned by General Motors, National Tank Company, Dresser Industries, and National Supply Company.[11] And towering above the entrance to the new building was the "Golden Driller," eight stories high and weighing 43,500 pounds, a gift of Mid Continent Supply Company and the permanent symbol of the IPE and Tulsa as the Oil Capitol of the World.[15]

Martin C. Dwyer, Incorporated, manager of the IPE, instituted new policies for exhibitors, a "package plan" that enabled them to reduce their expenses in readying displays for viewers and customers. Under this plan an exhibitor, when he paid fees for space, was relieved of the need to shop for local contractors to handle transportation and to set up his equipment on the exposition grounds. The IPE would hire the labor and provide all facilities. According to Dwyer, "All the exhibitor had to do was get the material to the exposition gates. The IPE management [took] over from there."[16] This brought exhibitor costs down to $5.50 per square foot, 10 percent lower than what exhibits had cost in 1959.[17]

Less popular but in accord with the wishes of exhibitors was segregating the general public from oil industry professionals. The first four days of the show were designated for the "general public," and the last six days were reserved for the guests and customers of exhibitors. "This is a common trade show practice," Robert Latta, assistant manager under Dwyer, said.[18] W.K. Warren explained, "Exhibitors have complained that business clients have been hampered by the great crowds that have visited the show throughout the entire period."[19] (There still is debate about this decision among IPE aficionados.)

In January of 1966 tradition was broken by election of the first woman to the IPE board of directors. Maxine Hacke, supervisor of records control for Warren Petroleum Corporation and a former international president of Desk & Derrick Clubs, became one of 143 board members. Also in January of 1966 IPE officers were reelected, and five members were added to the executive committee. W.K. Warren again served as president with Frank J. Hinderliter and W.H. Helmerich II as vice presidents. J.L. Shakely was secretary-treasurer, and Mrs. Virginia Davis was named assistant secretary-treasurer. Added to the executive committee were John L. Loftis, general manager of Humble Oil and Refining Company, central region; Joseph B. Kennedy, president of Sinclair Oil and Gas Company; Dean A. McGee,

The "Golden Driller" just days before the 1966 show opened. Courtesy Oil and Gas Journal.

chairman and president of Kerr-McGee Oil Industries, Inc.; J.C. Donnell II, president of Marathon Oil Company; and Claude B. Brownley III, Tulsa marketing division manager for Texaco.[20]

In March of 1965 congress passed a resolution, as it had done for previous IPE shows, authorizing the president to invite the "states of the union and foreign nations" to participate in the great extravaganza.[21] Planning then got underway for the show to be held slightly more than a year later.

On May 12, 1966, Andy Andrews (whose famous yell, "Let's Go-o-o-o Tulsa!" could be heard at every Tulsa Oiler baseball game) stepped to the microphone in front of the new IPE Building and opened the 15th IPE with his resounding yell.[22] Principal speaker at the opening ceremony was Texas Governor John Connally, who was joined on the platform by Governors Henry Bellmon of Oklahoma and William Avery of Kansas along with the heads of oil companies and state monopolies from foreign countries. Presiding over this ceremony was IPE President W.K. Warren.[24] In his remarks Governor Connally called for "federal and state governments to do everything possible to reverse the slump in exploratory drilling,"[25] calling for greater incentives for oil companies to explore for more oil. Little could he realize how prophetic his remarks would become in the decade ahead.

In his remarks President Warren reviewed the progress of the IPE across the years, pointing out that since that first show in 1923 the extravaganza had grown more scientific. He concluded "This 1966 exposition promises to be the best in our industry's history."[26] Tulsa Mayor James M. Hewgley officially opened the exposition at 12:32 p.m. by cutting a steel ribbon.

Inside the giant new building the exhibits revealed the transformation underway in the industry, a change wrought by automation. "With the exception of the aerospace industry," wrote Bob Foresman in an article in the *Tulsa Tribune*, "the oil industry has used automation to a greater extent than any other."[27] O.W. Graham, president of Instruments, Inc., a division of National Tank Company in Tulsa, predicted, "Within a few years the above-the-ground storage of oil will be all but eliminated."[28] Such a procedure already was standard in some fields in Africa and the Middle East where computers turned wells on and off as oil was needed. Graham added that the time soon would arrive when the computer would dictate how wells could be pumped to best advantage.[29] Another executive predicted that in the future customers would insert credit cards into gasoline pumps instead of paying a service station attendant.[30]

Riley Wilson, oil editor of the *Tulsa World*, found IBM's "light pen" to be "really amazing."[31] The pen's user would "squirt" the pen's light beam at points on a display screen and press a few buttons to draw graphs and charts, while the computer automatically stored these in its "brain." This system allowed users to draw maps and charts based on

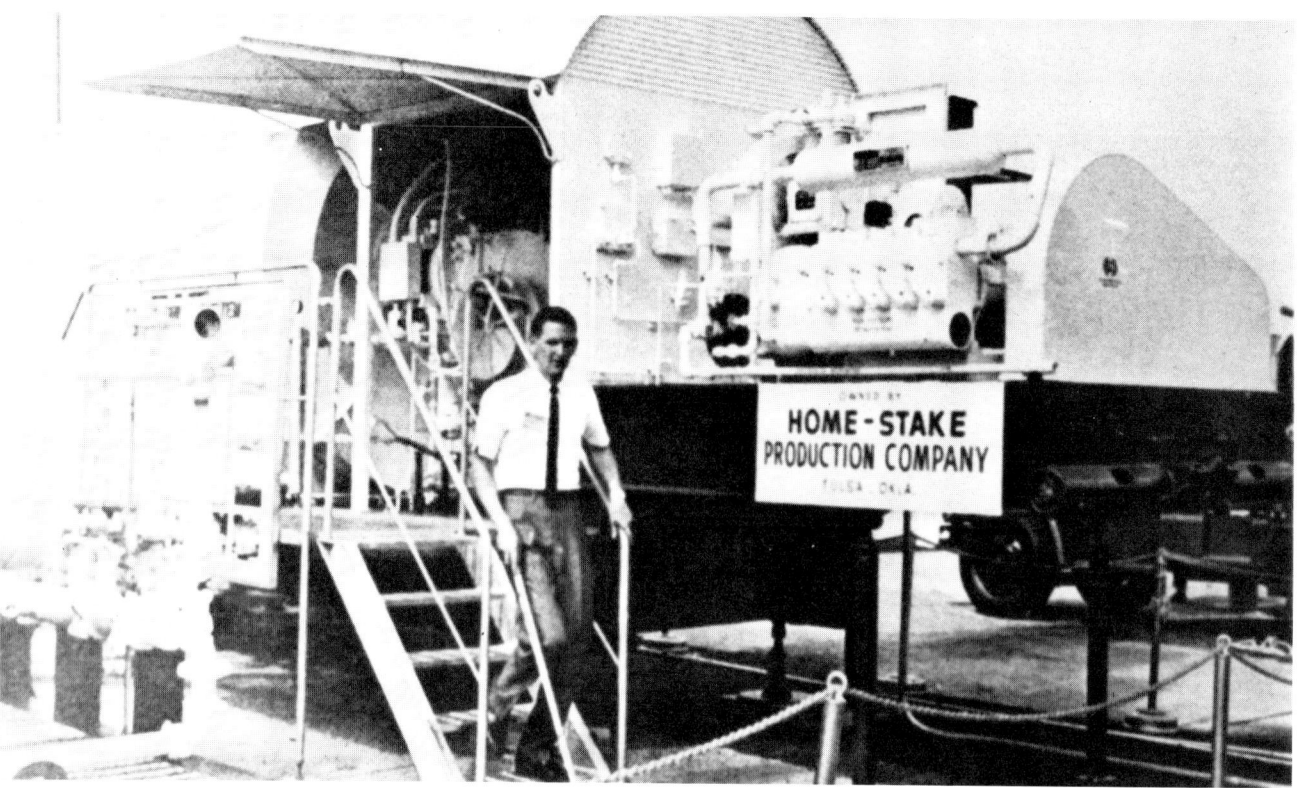

Home-Stake Production Company steam generator, displayed in 1966, was headed for duty in California. Courtesy Oil and Gas Journal.

A digital plotter displayed in 1966 by California Computer Corporation. Courtesy Oil and Gas Journal.

seismic data, contour maps, electric logs, decline curves, critical path diagrams, and other geophysical displays. This computer system, said Wilson, promised to "remove the drudgery of preparing drawings, graphs and charts" and thus would allow "technical men to spend virtually all of their effort in the analysis and evaluation of data."[32]

Computers already were being used to determine what was produced at a given refinery, what wells would be pumped on a given day, where wells should be drilled, and even the preparation of checks for stockholders and company employees. One firm, by consulting a computer, had saved $400,000 on the purchase of tubular goods, on which it spent some $4 million a year. K.M. Lawrence, vice president of Amerada, noted that "computers are used for storing vast reservoirs of geological data."[33] An "ultimate," he said, would be to use the machines to pinpoint drilling locations, a goal toward which some experiments were pointing.[34]

Also exhibited at the 15th IPE were traditional types of oil field equipment, likewise sporting advances. One of these was an experimental truck shown by the Chevrolet Motor Division of General Motors, the Turbo-Titan III. It had a gas turbine engine and featured "futuristic styling in fiberglass and steel with sophisticated control de-

A nuclear powered, acoustically operated subsea Christmas tree control system developed by Stewart, Stevenson, and Martin Company. Courtesy Oil and Gas Journal.

This work submarine, shown by North American Aviation in 1966, was designed for ocean-bottom work in offshore drilling and production. Courtesy Oil and Gas Journal.

At the 1966 IPE, CIBA Products Company demonstrated laying plastic pipe in a hurry using a Bell Helicopter. Pipe could be unreeled and strung by one man. Courtesy Oil and Gas Journal

vices and advanced features for safety, convenience and comfort."[35] The turbine engine, according to Chevrolet engineers, was quieter than conventional gas engines and had few moving parts, making it almost free of vibration and giving it the potential for "longer and more trouble free service."[36] The estimated life of the GT-309 Turbine was "at least" 350,000 miles.[37]

Also on display was Caterpillar Tractor Company's Cat 594, the world's largest pipelayer. This machine weighed 120,000 pounds and could lift 200,000 pounds. Four of these big "Cats" were in use on the $175 million Trans-Alpine Project, a 320-mile, 40-inch system to carry crude from tankers at Trieste, Italy, to refineries in Bavaria, Germany.[38] While the Cat 594 was climbing the Alps, North American Aviation's Beaver was plumbing the depths of the ocean. The Beaver was a submarine-workboat designed to complete wells 1000 feet deep in the ocean. The vessel's uses ranged from exploration and survey of the sea bottom to aiding in drilling operations and the installation of flow lines.[39]

Oil professionals and casual visitors alike were awed by the futuristic devices, processes, and machines on display, but such gagetry did not overshadow the presence of Myron Kinley. By 1966 he had quit fighting oil fires, surrendering his place to Paul "Red" Adair, who was becoming equally famous. Kinley was at the 15th IPE as a salesman for the downhole caliper manufactured by his son's company and for his own "expandable liner" for patching holes or corroded spots in downhole tubing.[40]

The great new IPE Building and the marvels it contained in 1966 were such a success that *Business Week* on May 21 devoted a two-page spread of text and pictures to it. The emphasis of the article was on the importance of automation to the oil industry. And IPE management took steps to assure that a proper record was made of the show in 1966. Tulsa's renowned photographer Howard Hopkins had a contract to serve as "official photographer" for the event, "one of the biggest photographic assignments ever given to a photographer in the Southwest."[41] Hopkins and his staff of 22 assistants shot and processed more than 5000 photos during the show, many of them to find their way into trade journals and newspapers around the world.[42] In addition, a documentary motion picture was made of the show for use as an historical record and for promotional efforts in signing exhibitors for the next IPE.[43]

This film chronicled not only the "marvels of machinery" displayed on the 1200 x 380-foot floor of the IPE Building, but also it showed the "people events" so much a part of the show. The Desk & Derrick Club's "First Lady of Petroleum" was Dr. Ivy M. Parker of Plantation Pipeline Company.[44] Selected by a panel of top industry executives, she was honored at a banquet on May 14 presided over by Mrs. Gracie Mitchel, Tulsa Desk & Derrick Club president. The principal speaker was George H. Weber, editor of the *Oil and Gas Journal*, and to her he gave an engraved scroll and a diamond wristwatch.[45]

On May 15 came "Old Timers Day," the 12th such event since its inception in 1927. Under the chairmanship of independent oil operator William J. Sherry, the Old Timers committee planned special ceremonies at 3:00 p.m. that Sunday to recognize pioneers with more than 40 years' service to the oil industry. Held near the entrance of the IPE Building, the ceremony honored 86-year-old "Pioneer of Pioneers" Van S. Welch of Artesia, New Mexico. The "Grand Old Men" honored included J. Howard Pew, 84 of Philadelphia, Pennsylvania, in refining; A.N. Hormol, 71, of Bernice, Oklahoma, in pipelines; John E. Elliott, 79, of Austin, Texas, in supplies and equipment; Frances Edgar Rice, 80, of Bartlesville, in natural gasoline; Paul Stock, 72, of Cody, Wyoming, in production; and Norris G. McGowen, 76, of Shreveport, Louisiana, in natural gas.[46]

When the show closed on May 21, the transition from old to new format was termed "pleasant."[47] The county fair atmosphere of the earlier period was gone, replaced by emphasis on business-customer relationships, and most participants found the transition less traumatic than expected. Total attendance was some 350,000 from all 50 states and 54 foreign counties, while those who came for the business-only portion numbered 143,287.[48] In a joint statement President Warren and Manager Dwyer noted, "We are pleased with the show because the exhibitors seem pleased with the contacts made with customers and prospective customers....From what exhibitors tell us [the show] was highly successful both in quality and quantity of oil men."[49]

A little publicized part of the show, which also seemed to please exhibitors, was an audit of attendance figures by a division of the Audit Bureau of Circulations, a Chicago-based organization, to ascertain whether claimed attendance figures were real or inflated. Manager Dwyer said that the auditors had gone about their task "in an impersonal, thorough manner" and that the data they collected would be "fed through an IBM system so that several categories of data [would] be made available." Apparently the result satisfied both the exhibitors and IPE directors, for when the IPE board met on June 14 to review reports of the successful show it announced that the 16th IPE would be held in 1970. However, this date was later moved forward one year.

1971

As 1971 dawned, thousands of independent operators believed that the United States was "on the threshhold of a major effort to increase its domestic exploration activities and producing capacity in view of the alarming fuel crisis now upon us."[51] Officials of the IPE decided that year to make a special effort to help these independent operators and producers as well as operators of small drilling and well-servicing companies. This decision was made after a survey by Leslie Brooks and Associates which indicated that there were some 5000 companies doing "exploration and production activities year-in and year out."[52] Of these 5000 firms, some 200 bought 50 percent of the services and supplies and thus received close attention from the sales staffs of equipment manufacturers. This left the other 4800 companies to fend for themselves in finding out about the latest and best in equipment and supplies. IPE officials therefore decided to remedy this situation by an intense campaign to attract these independent operators and producers to the IPE. This would be done mainly by recognizing the important contributions these people and their various associations made to the industry. However, the IPE would not lose its international flavor despite this new emphasis; it simply would add a new constituency largely neglected by the industry's manufacturers.[53]

Another feature of the 16th IPE was its emphasis on the various divisions of the oil industry. Most days of the show were set aside for recognizing one or two of these divisions. May 17 would be Exploration Drilling Day under the direction of committee chairman W.R. Gannaway of Sun Oil Company and Paul Lyons of Atlantic Richfield, along with Lee Daniels of Helmerich & Payne and E.A. Smith of Service Drilling Company. Activities on Production-Pipeline Day, May 18, were under the direction of H.S. Erskine of Kewanee Oil Company and J.L. Rogers of Humble Oil and Refining Company, Oklahoma City; they were assisted by N.B. Marvris of Continental Pipeline Company and Herold C. Price of H.C. Price Company. May 19 was Refining-Petrochemical Day, and its activities were coordinated by H.H. Belew of Skelly Oil Company, David L. Glaser of Champlin Petroleum Company, T.L. Cubbage of Phillips Petroleum, and Jack Roach of Kerr-McGee. May 20 was Offshore Day and was under the direction of Jack H. Abernathy of Big Chief Drilling Company and J.W. Bates, Jr., of Reading & Bates.[54]

Overall direction of the IPE in 1971 was the responsibility of its officers: F. Randolph Yost of Pan American Petroleum Corp., president; W.H. Helmerich II, chairman of Helmerich & Payne, vice president; E.B. Miller, Jr., president of Skelly Oil Company, vice president; J.L. Shakely, retired president of Jones & Laughlin Supply Company, secretary and treasurer; and P.C. Lauinger, Jr., president of Petroleum Publishing Company, assistant secretary-treasurer. Added to the executive committee as a replacement for G.D. Alman, Jr. (who remained on the general board of directors), was Dewey Bartlett, former governor of Oklahoma and former chairman of the Interstate Compact Commission. Serving with Bartlett were John A. Armstrong, D.D. Bovaird, E.F. Bullard, F. Allen Calvert, L.E. Cranston, J.C. Donnell II, J.E. Heston, John M. Houchin, Jack Judd, Robert L. Kidd, Dean A. McGee, Roy M. Hays, Robert Sharbaugh, and M.A. Wright. On the general board of directors of the IPE, Dewey Bartlett replaced his brother David, co-owner of Keener Oil Company of Tulsa. This group of executive officers was elected by the 145-member general board on January 22, 1971, and again the Chicago firm of Martin C. Dwyer, Inc., was hired to manage the IPE.[56]

Exhibitors long since had been signing up for space. Four days after the first application forms were mailed, registrations began pouring in, and improvements were undertaken on the IPE Building. Display space was expanded, utility outlets were more conveniently placed, the building was completely enclosed, and temperature controls were extended through the facility.[57] Everything was made ready for what President Randolph Yost expected to be "the greatest show ever from almost every standpoint."[58] Yost could make this statement because by December of 1970, five months before the show opened, 81 percent of the display space had been sold and $1 billion worth of equipment was scheduled for display inside the IPE Building and additional displays were planned for outside. IPE housing committee co-chairman F. Allen Calvert, Jr., and E.F. Bullard were expecting an overnight visitor population in excess of 30,000 on peak days of the show.[59]

A throng of enthusiastic oil men and Tulsans poured through the 21st Street entrance to the exposition grounds on May 16 for the opening ceremonies. Principal speaker was Rogers C.B. Morton, Secretary of the Interior, who was introduced by Senator Henry Bellmon. Morton's topic was the question of the Alaska pipeline and the controversy surrounding it.[60] "Ecology is the most wonderful thing in the world until the lights go out," he told his audience. "You have to look at it from a practical point of view." He concluded that exploration and production had to continue but with more concern for the land than had been the case in the past.[61] Others on the platform included Representative Page Belcher of Tulsa, Senator Clifford Hanson of Wyoming, and John Nassikas, chairman of the Federal Power Commission.

A new feature of the IPE in 1971 was a symposium series. According to President Yost, this was intended "to assist exhibitors in presenting in a more understandable and authentic way the vital contribution of oil equipment manufacturers to the advancing technology for developing and utilizing the world's petroleum resources."[62] IPE

Planning for the 1971 IPE are (L to R) Martin Dwyer, IPE Manager; Randolph Yost, president; and M.C. "Mike" Enright, a director. Courtesy Leslie Brooks.

directors expected the symposium to "awaken the interest of visiting oil men to the many readjustments resulting from the application of changes in an advancing technology."[63] As Yost noted, "There are readjustments required by deeper offshore locations and deeper drilling; by new regulations to control air and water pollution; by drilling and producing operations under extreme climatic conditions; and from the demand for trans-oceanic transportation of liquified natural gas."[64]

The symposium began on May 17 and continued for five days, covering such topics as: "Artic Operations," with panelists C.R. Steward, George Hughes, Jr., Ralph F. Cox, and P.N. Gammelgard; "Air, Water and Soil Pollution," with panelists W.B. Halliday, L.A. McReynolds, and P.N. Gammelgard; and "New Products and Recent Developments in the Petrochemical Industry," with panelists Dr. E.T. Guerrero, William C. Douce, and Charles B. Reeder. Thomas Baron spoke on "Today's Offshore Technology, Drilling and Production," followed by a panel discussion by I.H. Hughes, F.T. Pease, and Carroll C. Taylor; and "Liquification, Overseas Transportation, Storage and Gasification of LNG," with panelists David R. Williams, Bob L. Galloway, and Joe R. Wright. And there were panelists from Canada, Japan, and Europe who spoke at these sessions.[65]

The Ecology Hall of Science had an educational emphasis. This was provided by the IPE board of directors especially for the general public, which was invited only during the last four days of the show. Space in the Ecology Hall was given

Leslie Brooks, whose association with the IPE began in 1923 (when he attended as a Boy Scout, earning credit for a Civil Merit Badge). His advertising firm took over publicity for the IPE in 1934 and continued in that capacity until the show closed in 1979. Courtesy Leslie Brooks.

This rig, shown in 1971, was one of the largest trailer-mounted in the industry. Courtesy Oil and Gas Journal.

free to "qualified oil companies, oil trade associations, IPE exhibitors and educational institutions."[66] Displays were required to be non-commercial in nature and to show how the environment was being protected from the harmful effects of oil exploration, production, and use. As President Yost noted, this display was intended to show "the industry's approach and solutions to the problems of environmental pollution."[67] The Ecology Hall had some 45 exhibits, which were coordinated by former Oklahoma Governor Dewey Bartlett and George Roberts of Pan American Petroleum Corporation.

Another feature was the Ecology Film Festival, which consisted of 24 industry-sponsored films shown in the Exposition Center (near the Ecology Hall at the east end of the building). Two of the films, produced by the American Petroleum Institute, had their first Oklahoma showings at the festival.[70] These films were run continuously five hours each day in a theater seating more than 100. James O. Kemm, then of the Oklahoma Petroleum Council, directed this effort. At the time he told reporters, "Tell the people to come in, watch a film or two and rest their feet."[71]

Two of the most popular "regular" exhibits were Continental Emsco Corporation's "Robot Girl" and Halliburton's simulated trip down a 20,000-foot oil well.[72] The Robot Girl, operated by a real girl, gave a taped spiel promoting Continental Emsco's products, while Halliburton's trip gave "riders" an opportunity to go down with a drill. Another exhibit allowed viewers to watch divers work on a pipeline underwater; this was staged in a 126,000-gallon tank with glass windows. Crowds also were drawn to a display of turbine power plants described by Keith Griffis of the *Tulsa World* as the "up-and-coming power source for a wide range of petroleum industry applications."[73] Major suppliers of turbine power plants included Ford, the Solar Division of International Harvester, Garrett-Airesearch, the Detroit-Allison Division of General Motors, Waukesha, and Orenda, Limited, of Canada.[74] Other exhibits included a versatile twin-engine helicopter, shown by Bell Helicopter, and a "North Slope Boot" shown by the Red Wing Shoe Company. This boot featured a safety steel toe and rugged construction, and it had been tested on the North Slope of Alaska at temperatures as low as 64 degrees below zero.

From the point of view of the directors of the IPE, the biggest news was a 4000-square-foot display sponsored by the Council of British Manufacturers of Petroleum Equipment for the British Board of Trade. Manager Dwyer noted that this was the first exhibit by a sizeable number of British manufacturers at the IPE. Represented in the display were 14 British manufactuers.[76] Other foreign exhibitors included firms from Canada, Argentina, Venezuela, France, Mexico, Russia, and Brazil. Visitors from these and other countries again were guests of the NOMADS IPE committee, co-chaired in 1971 by Don Collins of A-1 Bit & Tool Company of Tulsa and Roger LaPlante, sales representative for *Societe Entrepose Gtm Pour Les Trebaux Petroliers Martimes* of Paris. As early as December of 1970, NOMADS members had begun delivering engraved invitations from Oklahoma Governor Dewey Bartlett and IPE President Randolph Yost to embassies in Washington, D.C.[77] After authorization by a joint resolution of congress, President Richard Nixon also had invited foreign delegates to visit and participate in the IPE.

William Gallman, president of the Tulsa-Oklahoma City NOMADS chapter, and Harry E. Estes, executive secretary of the national NOMADS board of regents, announced in January of 1970 that the organization again would undertake hosting duties at the IPE, and in 1971 an effort would be made to give the affair a touch of elegance. A fleet of 15 white Cadillacs were placed on special call to transport foreign delegates; these arrangements were made by G.W. Davidson, president of W.C. Norris Division of Dover Corporation. On May 20, these new white Cadillacs pulled up to the Tulsa Assembly Center to deliver their passengers to a great international party planned and directed by NOMADS' Roger LaPlante.[78]

Foreign visitors were left wanting for nothing in 1971. Their registration, housing, and entertainment had all been minutely coordinated by NOMADS with assistance from the Tulsa Desk & Derrick Club; D & D President Mary Hembree of Atlantic Richfield gave great assistance. An air conditioned lounge was provided for foreign delegates along with a message service, help in translation, transportation, a business-liaison service, directory service, and any other form of assistance needed. The NOMADS and Desk & Derrick Club outdid themselves to assure that the IPE was a hit with international visitors.[79]

The women of Desk & Derrick also prepared to celebrate the selection of their fourth "First Lady of Petroleum." Chairman for this event was Ruth Hamill, who announced that in 1971 D & D clubs also were celebrating their 20th anniversary.[80] The judges that year selected another scientist, Dr. Margaret Boos, a geologist from Denver, Colorado, to receive the coveted honor.[81] The award was made at a banquet on May 22 in the Crystal Ballroom of the Fairmont Mayo Hotel. Special guests included Dr. Ivy Parker, the "First Lady" of 1966, and Inez Autey Schaeffer of Tulsa, the founder of the first Desk & Derrick Club. Featured speaker at the event was Retha Odom, manager of public relations for the research and development division of Shell Oil Company. The official hostess at the banquet for the Tulsa D & D Club was Patricia Henderson of Amoco Production Company. Dr. Boos was presented with a scroll and a diamond wristwatch.[82]

By this time the IPE Old Timers committee, chaired by William J. Sherry, had handed out its honorary titles. John W. "Jay" Bird, 89, of Bradford, Pennsylvania, was named "Pioneer of Pioneers," and "Grand Old Men" were named in each major division of the oil industry: Cyrus McDonald Scott, Jr., 80, of Tulsa in pipeline/transportation; I.A. O'Shaughnessy, 85, of St. Paul, Minnesota, in refining; Daniel Pyzel, 94, of Piedmont, California, in petrochemicals; Robert W. Hendee, 76, of Colorado Springs, Colorado, in natural gas; Stanley B. Crooks, 81, of Bartlesville, in gas processing; Glen J. Smith, 86, of Tulsa, as landman; Frank Rinker Clark, 90, of Tulsa, in exploration; Joseph Zeppa, 77, of Tyler, Texas, in drilling; J.P. Owen, 73, of Lafayette, Louisiana, in production; and William L. "Bill" James, "retired," in purchasing. These awards were presented at a banquet at which P.C. Lauinger, publisher of the *Oil and Gas Journal* served as master of ceremonies, and IPE President Yost made the presentations.[83]

The IPE itself received an award in 1971, given by the Oklahoma Petroleum Council and the Oklahoma Historical Society in the form of a seven-foot-high granite monument erected at the Exposition Center's main (south) entrance. Dedicatory ceremonies were held on May 16 under the direction of John Steiger of Cities Service Oil Company, chairman of the Petroleum Council's historical committee; G.R. Grainard of Atlantic Richfield, president of the Oklahoma Petroleum Council; and Elmer L. Fraker, executive director of the Oklahoma Historical Society.[84] The text on the marker noted that the IPE had started in Tulsa in 1923 and had grown from displays of $10,000 worth of equipment to exhibits valued at more than $1 billion.[85] The monument gave eloquent tribute to the 56-year history of the IPE.

The closing day of the 16th IPE, May 23, 1971, was dark and soggy, a condition to which the IPE faithful had become accustomed, but thousands still attended. Officials concluded that the effort in 1971 was not the largest IPE, but nevertheless it was highly successful. What pleased IPE officials most was the computer-confirmed attendance of 29,904 oil men (not including their wives and children). Moreover, there had been representatives from 62 foreign countries, prompting President Yost to call this "the greatest international spread ever recorded for the show."[86] He also noted that neither IPE officials nor exhibitors were disappointed at the lower attendance figures inasmuch as those who did come tended to be buyers. The lower attendance, Yost said, "is in line with the policy of many companies— including my own—to cut travel expenses by limiting exposition and convention attendance to department heads or those approximating that level."[87]

In summary, President Yost added, "After assessing the reaction of industry participants to this year's event, [the executive directors] have decided to begin immediately with plans for the seventeenth IPE."[88]

1976

When the 17th IPE opened in Tulsa in 1976, a major topic of conversation was a "hole in the ground in southeastern Oklahoma deep enough to dump twenty-four Empire State Buildings down…a hole more than 5.9 miles straight down."[89] Most people remembered that both the equipment and the drill making this hole had come from manufacturers who had unveiled their prototypes "at the greatest of all oil and gas trade shows," Tulsa's IPE.[90]

Streamlined to just five days' length and labeled "Energy '76," the 17th IPE was the result of hard work by an executive committee composed of Randolph Yost, W.H. Helmerich II, M.C. "Mike" Enright, Phillip C. Lauinger, and Ernest B. Miller, Jr. They received strong assistance from the other executive directors: John A. Armstrong, Dewey F. Bartlett, D.D. Bovaird, E.F. Bullard, F. Allen Calvert, G.P. Bunn, Jr., J.C. Donnell II, E.P. Hardin, John M. Houchin, Warren L. Jensen, Jack Judd, Dean A. McGee; Fred H. Ramseur, Jr.; H. Robert Sharbaugh; and M.A. Wright. Tulsa Mayor Robert LaFortune declared the week of May 17-21 to be International Petroleum Exposition Week in Tulsa. The 17th IPE was actually two shows in one: the main portion of the show arena was for the petroleum industry, but a 36,800-square-foot area had been set aside for the general public. This housed exhibits in the "Energy Science Panorama," a collection of non-commercial exhibits covering all sources of energy and featuring an "Energy Film Festival." This Panorama, said President Yost, had been created by public demand and for the first time made the IPE truly a total energy show. Although the public was not encouraged to tour that part of the IPE set aside for the petroleum industry, visitors without industry exhibitor invitations could register at the south entrance of the Exposition Center, pay a fee of two dollars, and see the entire show.

Displays in the Panorama varied from a "donkey-powered threshing machine and windmill" to "solar panels and energy crystals"[93] and constituted the largest accumulation of energy products, equipment, and technology ever shown in one display. Exhibitors ranged from private individuals showing their creations under auspicies of the Oklahoma Inventors' Congress to the United States Energy Research and Development Administration, which sponsored the Federal Energy Center, a modular exhibit featuring "American Energy—the Challenges Ahead."[94] At the Federal Energy Center, movies were shown about coal gasification, and there was a three-stage gasifier. In addition, the U.S. Geological Survey demonstrated its "Helium Sniffer," which was used to find geothermal uranium and oil reserves, while the Corps of Engineers demonstrated how hydroelectric energy was generated.[95]

Also in the Panorama was a "university row" which included the participation of nine universities: Massachusetts Institute of Technology, Rice University, Colorado School of Mines, Missouri School of Mines (Rolla), the University of Wyo-

This granite monument was erected at the entrance to the Tulsa Exposition Center to commemorate the work of the IPE. It was donated by the Oklahoma Historical Society and the Oklahoma Petroleum Council. Looking on are John Steiger (left) and R.L. Kidd. Courtesy James O. Kemm.

ming, the University of Tulsa, Oklahoma State University, Princeton University, and the University of Oklahoma.[96] Oil companies likewise participated in the Panorama section of the show.

Cities Service entered an operating model of the micellar-polymer flooding system used for tertiary recovery of oil from old wells. Combustion Engineering contributed an exhibit on waste con-

Space for the 1976 show was 52 percent sold out 12 months before the show. Robert Sharbaugh (left), president of Sun Oil Company and an IPE director, discusses the upcoming show with William J. Sherry, an independent producer and long-time chairman of the Old Timers Committee. Courtesy Oil and Gas Journal.

version, while McDonnell-Douglas showed how space age technology was used to construct thermopiles for the trans-Alaska pipeline. The Batelle Company had a display dealing with nuclear power, while Southwest Research Institute showed remote-control equipment for handling materials in hazardous areas. Other companies exhibiting in the Panorama included IBM, Public Service Company of Oklahoma, Kerr McGee, Phillips, the Alyeska Corporation, Owens-Corning Fiberglass, Amoco, Bechtel, and Braden Industries.[97]

According to George Howard of Amoco, one government official expressed surprise that "the highly competitive oil industry would be willing to share its technology and achievements." Howard told him "that we also welcome press covereage and if we have a collective disagreement perhaps we can come up with a solution."[98] He also explained that almost two years had been required to select the exhibits, all of which were shown at the expense of the exhibitors. The Panorama, which was open to the public without charge, was designed to explain in non-technical terms the various facets of energy development, proposed innovations, and the potential of exotic sources of energy. Included in it was a continuous showing of energy-related films.[99]

Energy '76 was more than a "visual display of the nuts and bolts of the oil and gas industry."[100] A six-session, three-day symposium sponsored by NOMADS focused on possible methods of recovering "billions of barrels of oil already discovered but not recoverable by conventional methods."[101] The first session of this symposium opened at 9:00 a.m. on May 17 in the IPE Educational Building (north of Exposition Center) and was titled "Trends in Equipment for Exploration of Hydrocarbons, Uranium and Geothermal Energy." The papers were presented under the chairmanship of Dr. E.T.Guerrero, Dean of Engineering and Physical Sciences at the University of Tulsa. Moderator was H.J. Kidder, a planning analyst for Mobil Exploration Service Center, Dallas, and panelists included Carl H. Savit, I.A.P. Rumsey, Julian A. Pawley, C.H. Raitz, George V. Keller, and M.P. Tixier.

The topic of session two was "Drilling for Oil and Gas in Hostile Environments (Deserts, Snow, Tundra, Great Depths, Etc.)." Moderator was Charles E. Thornton, president and chief operating officer of Reading & Bates Offshore Drilling Company of Tulsa; speakers included A. Doug Brown, Neal V. Perry, W. Thomas Adams, J.N. Muscovalley, A.R. McLerran, and Max J. Fritzel.[103] Session three, held on May 18, was moderated by Dr. Kermit E. Brown, Halliburton Professor and Chairman of the Division of Resources Engineering at the University of Tulsa; speakers included T.M. Geffen, G.C. Lahn, B.B. Fuqua, D.E. Wittig, L.R. Peddycoart, R.E. Hurst, Gene Snell, and S. Donald Moore.[104] The afternoon portion of session three covered the topic, "Improvements in the Field of Petrochemical Processing of Hydrocarbons," and was moderated by Dr. C. Vernon Foster of Continental Oil Company, Ponca City; speakers included Willis J. Service, Donald Ethington, R.T. Schneider, James A. Mitchell, Art Corrigan, Glenn E. Handwerk, Turner C. Smith, Jr., Paul H. Washecheck, Ronald L. Poe, Terrence B. Kryska, and Stanley S. Paist.[105]

On the last day of the symposium, May 19, the subject was "Developments in the Transportation and Storage of Hydrocarbons." Moderator for the day was Morgan Greenwood, president of Resource Sciences Corporation of Tulsa. Panelists included Dr. Abdulhady H. Taher, Robert W. Ward, Joseph B. Harnett, James Donald, and Dr. William A. Fisher. That afternoon the symposium concluded with "Development and Potential of Future Energy Sources—Coal, Shale Oil, Nuclear, Solar, Geothermal," moderated by Dean A.

Austin N. Heller, Assistant Administrator for Conservation, U.S. Energy Research and Development Administration, speaks at the "Engineers' Day" Luncheon on May 19, 1976. Courtesy Leslie Brooks.

McGee, chairman of the board of Kerr McGee of Oklahoma City. The speakers were Dr. Chauncey Starr, Dr. Henry R. Linden, Raymond L. Dickeman, Harry Pforzheimer, Dr. Carel Otte, Erich H. Reichl, and Michael C. Noland.[107]

The exhibits at Energy '76 were heavily onshore-oriented at a time when offshore work was becoming the "glamor" segment of the industry. However, 92 percent of all operating rigs were onshore, and onshore activity accounted for a majority of exploration and production expenditures. Several pieces of new equipment were unveiled at the IPE, for this was still the "big daddy" of oil shows "anyway you look at it."[108] Charles R. Morse, Inc., of Anchorage, showed its "Pipehandler," used to place the 38-inch-diameter pipe on the Alaskan Pipeline project. Another interesting item was the "Hydroport" shown by Dickson Brothers, Inc., of Tulsa; this was a portable water system capable of producing drinking water in the field anywhere in the world.[109] From Canadian companies came exhibits on the "Quadrem Multifrequency Airborne Electromagnetic System" and the "Hydrosonde Deeptow Seismic System"; both of these Canadian exhibitors were part of a group called Venturtek International, Ltd., of Toronto, while two other Canadian firms displayed geophysical equipment.[110]

Norwalk-Turbo, Inc., a subsidiary of the Union Corporation, exhibited a new turbo-compressor package, the TC-7, said to be the smallest and lightest such unit then available.[111] New type

The NOMADS

cordially extend an invitation

to

to attend the

International Petroleum Exposition and Congress

May 17 through 21, 1976
Tulsa, Oklahoma, U.S.A.

and

to enjoy the courtesies of the

NATIONAL OIL EQUIPMENT MANUFACTURERS
AND DELEGATES SOCIETY

register at Nomads Lounge, Exposition Center, Tulsa

I most sincerely hope you can make it!

An official invitation to foreign delegates in 1976 from NOMADS. Courtesy Leslie Brooks.

Past Desk & Derrick Club presidents honored in 1976. L to R: Floriene Messall, 1974 president; Frances Hidell, 1972; Gloria Caravantes, 1975; Lilly Wright, 1976; and Mary Lee Turner, 1973. Coutesy Tulsa Daily World.

pressure gauges, flow meters, and moisture analyzers were unveiled by various manufacturers,¹¹² while the Alemite & Instrument Division of Stewart-Warner displayed a "portable service station," a truck mounted facility designed to bring complete lubrication service to all types of equipment in the field.¹¹³ Other exhibitors included E.I. duPont de Nemours & Company, Inc., H.O. Trerice Company, Geophysical Research Corporation, Basic Science, Inc., Geograph Company, and C-E Natco (formerly National Tank Company). Bob Foresman of the *Tulsa Tribune* described the water filtration system shown by C-E Natco as "breath-taking," a description that could be applied to the entire display area at the 1976 show (for which all space was sold a year before the show opened).¹¹⁴

Other features of the IPE continued basically unchanged despite the new five-day format. NOMADS still poured its energy and ingenuity into preparing a spectacular reception for foreign delegates. That year NOMADS doubled its space reservation for the International Delegates Lounge and provided a cadre of interpreters from the University of Tulsa to assist speakers of all languages. Herbert C. Fries, general chairman of the NOMADS international committee, and his fellow workers tried to anticipate every problem foreign delegates might face and be prepared with solutions.¹¹⁵ As usual since 1950, NOMADS had the enthusiastic assistance of the Desk & Derrick Clubs.¹¹⁶

The women of D & D also assisted the IPE with tours, registration, information services, and entertainment. For 26 years they had been helping where needed, and the IPE directors recognized the value of this work by setting aside May 21 as "Desk & Derrick Day" at the show. Events planned for that day included a reception and luncheon at the Mayo Hotel, during which past presidents of D & D were singled out for recognition. This replaced the customary "First Lady of Petroleum" contest. Honored were Frances C. Hidell, president in 1972, Mary Lee Turner, 1973, Flo Messall, 1974, Gloria Caravantes, 1975, and Lilly E. Wright, 1976. The ceremonies that day were coordinated by the activities committee of the Tulsa Desk & Derrick Club chaired by Bettye Cunningham, of the Franklin Supply Company of Tulsa, and Julia McCormick, of Agrico Chemical Company.¹¹⁷

Unchanged in 1976 was awarding of "Grand Old Man" titles to pioneers of the oil industry. In 1976 11 such men were honored, along with William J. Sherry, an independent operator from Tulsa for whom a special title was created. Sherry, who had been the driving force behind the Old Timers organization since 1953 and who had been involved in every IPE since the first one in 1923, was cited for "Distinguished Service to Oil Industry Pioneers." Sherry had come to Tulsa in 1921 and had joined Frank B. Murtan and Marshall Strozier in oil developments in Creek County. Strozier, 94 years old in 1976, said that the Carter Oil Company had backed the ventures of Sherry's group because it "liked everything Bill suggested and knew he would tell the truth before drilling."[118] Strozier added about Sherry, "He [was] one of the best thinkers I have ever had dealings with."[119] P.C. Lauinger added, "He is a good friend and a fine gentleman,"[120] while IPE President Yost spoke admiringly about him when presenting the award on May 17.[121]

"Grand Old Man" awards went to: George W. Clarke, 76, for landmen; John J. Moran, 79, for drilling; Charles Wheatley, 83, for pipeline/transportation; J.B. Saunders, 74, for refining; William K. Warren (former IPE president), 78, for natural gasoline; George P. Bunn, Sr., 83, for natural gas processing; Ross W. Thomas, 77, for petrochemicals; Dee W. Simms, 70, for purchasing; C.I. Wall, 72, for natural gas; and Wallace E. Pratt, 91, for exploration. Named "Pioneer of Pioneers" was J. Paul Getty, who began his career as a roustabout in 1909 with Minnehoma Oil Company in Osage County, Oklahoma.[122] Getty was unable to attend, but sent a telegram expressing his heartfelt thanks for the award and voicing his optimism for the future of the oil industry.[123]

Other members of the oil fraternity apparently shared Getty's optimism, for they swelled attendance figures at the IPE in 1976 to record levels for oil industry professionals. The five-day show drew 44,699 qualified buyers.[124] By Friday, May 21, the last day of the show in 1976, the IPE's executive directors already were busy planning a show for 1979 which would, they said, begin a new series of shows to be held at three-year intervals instead of every five years.[125]

1979

Before the opening of the IPE in 1979, a survey was conducted among the top people in the petroleum industry by Leslie Brooks & Associates to determine if the traditional constituency of the

L to R: Former presidents W.K. Warren and Randolph Yost congratulate John Houchin on his appointment as president of the 1979 IPE. Courtesy Leslie Brooks.

IPE would send exhibits and representatives to an oil show specializing in onshore operations. The title of Brooks' report was encouraging: "IPE Rates 'Tops' as Onshore Equipment Show Majors and Independents Prefer to Send Buyers/Specifiers To." It revealed that 77 percent of the respondents would send buyers and specifiers to an onshore show, that 68 percent of oil company equipment buyers/specifiers preferred an onshore show, and that 45 percent said the IPE was "best for them" as an exhibitor of onshore equipment. In 1976, 62 percent had sent representatives to the IPE; 56 percent of the major companies had expressed a preference for onshore over offshore expositions; and 77 percent of the independents had felt the same way. Although such results were encouraging, they reflected a basic change in the IPE toward specialization in one type of oil industry operation. The title of the 18th IPE announced this change: 'Energy '79: International Exposition and Congress: International Onshore Equipment and Service Oil Show."[127] IPE officials believed they were playing to the strength of the Tulsa show with this narrowing of focus, and the IPE faithful marshalled their forces for an opening on September 10.

After the show in 1976, there had been an important change in leadership. Randolph Yost had stepped down as president to return to the roster of executive directors. He was replaced by James E. Hara, president of Skelly Oil Company. However, Hara died before preparations were complete for the show in 1979, and John M. Houchin was named to the top position. The retired chairman of the board of Phillips Petroleum Company, Houchin had been a member of the IPE board of directors for many years.[128] Helping him lead the IPE were the other executive directors: W.H. Helmerich II, Dean A. McGee, M.C. "Mike" Enright, P.C. Lauinger, Jr., E.R. Albert, Jr., John A. Armstrong, Bob Diggs Brown, William C. Douce, Dr. Warren L. Jensen, Jack D. Jones, Jack Judd, Calvin McKee, Fred H. Ramseur, Jr., R.E. Sampson, H. Robert Sharbaugh, Randolph Yost, and Henry H. Zarrow.[129]

In April of 1979 President Houchin told the executive committee that the planned "onshore show has won favorable comment throughout the industry and among exhibitors."[130] He added that onshore specialization would "embrace 65 percent of the drilling and production budgets, 90 percent of running rigs, and more than 90 percent of producing wells and 98 percent of the pipelines."[131] However, as Riley Wilson noted in the *Tulsa World* of August 22, 1979, designating the IPE as an "onshore" show was an admission that the Tulsa extravaganza had lost some of its stature to the Offshore Technology Conference held each year in Houston: "Until this year the IPE had officially ignored the OTC although it has mushroomed in the last few years as offshore petroleum activity spread around the world."[132]

In addition, the IPE was cut to four days (September 10-13), its shortest run ever.[133] This was in line with what was happening with other trade shows.[133] In fact, there were so many trade shows that there was competition for the most desirable dates. Nevertheless, early registrations showed that the IPE had continuing support in the oil community. Publicity came at the local, state, and national levels. Three trade publications—*American Oil and Gas Reporter*, *Drilling Magazine*, and *Oil Daily*—announced in June that their next issues would spotlight the IPE.[134] And, as usual, the *Oil and Gas Journal* produced a special issue for the

JOINT RESOLUTION

Authorizing the President to invite the States of the Union and foreign nations to participate in the International Petroleum Exposition to be held at Tulsa, Oklahoma, from September 10, 1979, through September 13, 1979.

Resolved by the Senate and House of Representatives of the United States of America in Congress assembled, That the President of the United States is authorized and requested to invite by proclamation, or in such other manner as he may deem proper, the States of the Union and foreign nations to participate in the International Petroleum Exposition to be held at Tulsa, Oklahoma, from September 10, 1979, through September 13, 1979, for the purpose of exhibiting machinery, equipment, supplies, and other products used in the production and marketing of oil and gas, and bringing together buyers and sellers for the promotion of foreign and domestic trade and commerce in such products.

Joint Resolution of congress inviting nations of the world to the 1979 Onshore IPE. Courtesy John Houchin.

IPE. In addition, the U.S. Department of Commerce distributed 8500 two-color, five-language brochures about the IPE at 152 American embassies and consulates. Damon Greer of the Department of Commerce noted that there was a heavy demand for these brochures because of a worldwide concern over the "tightening energy situation and the determination of many nations to broaden their knowledge of the oil and gas producing industry."[135] Inasmuch as independent operators drilled 85 percent of the wells in the United States, most of them onshore, much of this publicity was directed at them.[136]

The IPE of 1979, like those immediately preceding it, was closed to the general public. Some participants and observers, such as James O. Kemm of the Oklahoma Petroleum Council, thought this policy ill-advised; they felt that public support for the oil industry could be generated by admitting everyone to the IPE and that such support was as valuable to the industry as a whole as were sales by exhibitors.[137] However, most exhibitors felt that milling throngs of sightseers interfered with the conduct of business at exhibit booths. There was, as usual, some accommodation to allow public participation in the show.

By August 22, 1979, the success of the show was assured with 85 percent of available space in the 10.5-acre Exposition Center under contract. This exactly matched the percentage of wells drilled by independent operators, although virtually all the majors were represented.[138]

On opening day, according to Bill Sansing of the *Tusla World*, oil men "like a swarm of hungry ants...packed the Onshore International Petroleum Exposition." They were "asking questions and examining the myriad technological developments in their field since the last show three years ago." And while oil men were thus engaged, "the general public was visiting the Hall of Science learning the latest about enhanced oil recovery."[139]

The exhibits at the Hall of Science focused mainly on experimental techniques of secondary, tertiary, and even more advanced recovery coming from some 75 experimental projects on which the federal government and American oil companies had cooperated.[140] John S. Ball, retired director of the Bartlesville Research Center for the Department of Energy, was director of the Hall, and coordinating the program was Fred H. Ramseur, Jr., executive vice president of Cities Service Oil Company of Tulsa. Others on the committee included Louis Alexander, Charles L. Coffman, R.C. Earlougher, Lloyd E. Elkins, M. Geffen, Joe R. Lindley, Maurice Miller, Riley B. Needham, and R.E. Sampson. President Houchin commented about the 20,000-square-feet in this portion of the exposition, "We feel this will be a big factor in interesting oil men to attend the [IPE], as they will get to see firsthand the progress that has been made in enhanced recovery, along with its possibilities for getting more out of the reserves that exist."[141]

Among the many displays were a game from Cities Service demonstrating the risks of enhanced recovery; a Getty Oil Company model of a steam flood operation (generator and pump); a Gulf Oil Company model of a carbon dioxide flood; various forms of enhanced recovery displayed by the Department of Energy; a Sandia Laboratories downhole steam generator with applications for petroleum recovery; and a Halliburton display of services to improve oil recovery. Fred Ramseur noted the potential demonstrated in the Enhanced Recovery Hall of Science by

The Mack Truck exhibit in 1979. The "slot" machines dispensed stuffed bulldogs, the Mack mascot. Winners had to get three Mack-related pictures in a row. Losers received a apiece of jewelry with the Mack logo. Courtesy Tulsa World.

commenting, "With continued technology improvements through experimentation, the oil industry will have many millions of barrels of oil recoverable from the existing reserves."[142] Supplementing the exhibits in the hall was a film presentation on the same subject sponsored by the Oklahoma Petroleum Council, with James O. Kemm overseeing the effort.[143]

Another program, sponsored by the Engineers' Society of Tulsa and chaired by John Zink, president of Tulsa's John Zink Company, related to research and development in the petroleum industry. This was a "productivity seminar" at which John Maher, executive vice president of the Los Angeles firm of Blythe Eastman Dillion & Company, made the keynote address (instead of Secretary of the Treasury William Simon, who was unable to attend). Maher noted, "The petroleum industry is in particular need of raising capital for research and development....There is plenty of room for innovation in everything from conservation to ventures into synthetic fuels."[144]

Other speakers and moderators included Jim Kilmer, Henry Keplinger, Herman Fritschen, Phil Wood, Ed Drake, Dr. Murray McComas, Dr. Larry Millinger, Dr. Eugene Swearingen, Bob Matthews, Jim Patton, Bill Thomas, David Orr, Carla Odell, Ira Gregerman, Bob McHardy, and Paul Anderson. Chairman John Zink told the thousands of engineers who attended the seminar that its purpose was "to generate national action through a nucleus of informed engineers to 'up' the United States' declining productivity [and to]

NOMADS and Desk & Derrick members plan for the 1979 IPE. L to R: Roy McGrann, Mary Smith, Herbert C. Fries, Francie Reagan, Ed Pelster, Charlotte Vieth, and M.C. "Mike" Enright. Courtesy Herbert C. Fries.

create more research and development funds and capital...."[146]

The IPE's own symposium dealt with several facets of the petroleum industry: "Equipment and Future Trends for Exploration of Energy Resources"; "Developments to Meet Current Drilling Challenges"; "Advancements in Production of Oil and Gas"; "Enhanced Oil Recovery Methods—Potential and Economic Feasibilty"; "Trends in the Transportation and Storage of Hydrocarbons"; "New Energy Sources—Potential and Development"; and "Politics of Energy Resources."[147] This symposium was chaired by Dr. E.T. Guerrero, former dean of the College of Engineering at the University of Tulsa. Moderators for these panels included Michael R. Waller, King P. Kirchner, L.M. Richards, John S. Ball, Vernon T. Jones, M.L. Sharrah, and Bill M. Burks.[148]

Although speeches and panel discussions had replaced the old fashioned ballyhoo of early expositions, nothing could replace the displays of oil field hardware, the main feature of every IPE. In 1979 these displays ranged from a power generation unit, the "Hole Saver," to a geological scale model of the Gulf Coast Salt Basin. Other items on display included vibration instruments, minicomputers, fire fighting equipment, winches of all sizes, metal springs, executive jets, pumping units which reduced gear box torque, pressure gauges, and meters of all types.[149]

Another IPE tradition of 40 years' length was the leadership of the NOMADS organization in promoting and hosting foreign delegates. Chaired by Herbert Fries, of Portable Gasoline Plants, Inc., of Tulsa, and assisted by Dr. E.T. Guerrero, who chaired the NOMADS symposium, the club again made foreign delegates welcome. The Tulsa chapter president in 1979 was B.W. "Bernie" Bennett who noted that the NOMADS had started the "presentations of technical symposiums at the IPE."[150]

And the Desk & Derrick Club of Tulsa, as it had for almost 30 years, played an integral part in making IPE programs and activities run smoothly. D & D members staffed the information booth at Tulsa International Airport, registered and typed name badges for all NOMADS and delegates at the show, registered visitors and provided information at the Old Timers Club booth, and assisted at the Oil City Center Lounge and at the Independent Oilman's Lounge. NOMADS and the IPE hosted a reception honoring all Tulsa and visiting D & D members for their contributions to the IPE. President of the Tulsa D & D Club in 1979 was Francie Reagan, of the Hughes Tool Company in Tulsa; Julia McCormick of Agrico Chemical Company was regional director; and Mary Elizabeth Pav, of the Ethyl Corporation at Baton Rouge, Louisiana, was national president.[151]

Eleven pioneers were honored at special ceremonies in the IPE Theater on September 10. Heading the list was "Pioneer of Pioneers" William T. Payne of Oklahoma City, president of Payne Petroleum and Seneca Oil Company. Named "World Pioneer" was Bart W. Gillespie of Fullerton, California; retired in 1964 as vice president of Home Oil Company, Ltd., of Calgary, he had been a geologist in Peru, Argentina, Mexico, and Canada. Others honored in 1979 included "Grand Old Men" in several categories: Ira H. Dram of Austin, Texas, exploration; Jonathan G. "Jack" Pew of Dallas, production; Otha H. Grimes of Tulsa, gas processing; Mark V. Burlingame of Bradenton, Florida, natural gas; Arthur O. Olson of Tulsa, drilling; Reid Brazell of Frankfort, Michigan, refining; Ben D. "Tex" Leuty of Atlanta, Georgia, pipelines; O.V. "Otz" Tracy of Morristown, New Jersey, petrochemicals; and Charles J. George of Odessa, Texas, well servicing.[152] This group averaged 56 years of service to the petroleum industry, and they were being recognized on the 56th anniversary of the IPE.

Few in attendance in 1979 felt disappointed when the show closed, for in almost every respect it seemed to have been a huge success.

EPILOGUE

When the IPE ended in 1979, there was considerable optimism about the future of the show, although attendance at the four-day event was below the 40,000 expected. The record showed that 32,318 had attended plus uncounted thousands at the Hall of Science, which was open free of charge to the general public.[1] President John Houchin announced that the board of directors would meet on the Thursday following the close of the show to vote on holding another show in September of 1982, a vote that was expected to be affirmative. Several exhibitors placed advance reservations for space in 1982, and the 27 foreign delegates who had registered expressed their intent to return in three years.[2] There had been a decline in attendance since the highly successful show in 1966, but there seemed no reason to doubt that the IPE had a future.

In November of 1979, however, Martin Dwyer, IPE general manager since 1963, submitted his resignation and suggested that the show be moved to Los Angeles, Denver, Dallas, or Chicago in order to spark greater interest. "In 1976," Dwyer noted in his letter to IPE director Michael Enright, "for the first time in IPE history, the exposition lost money and had to fall back on reserve funds. Most of the bills are now in for the 1979 show, and it is evident we will again have to fall back on reserve funds."[3] Dwyer also noted that the records for 1979 showed that only 5000 of those attending were from outside Oklahoma, "indicating that interest in the show from outside the state" had diminished. "Most shows," he continued, "rotate around and get a certain amount of enthusiasm when they go someplace they haven't been before."[4]

Director Enright countered Dwyer's comments by saying, "The International Petroleum Exposition definitely will be held in Tulsa in 1982, probably in September, with emphasis on the onshore operation of the industry,"[5] adding, "These newspaper reports [of the show's closing] appear to me to be much ado about nothing....The IPE originated by Tulsa merchants in 1923 always has reflected the condition of the oil industry. Right

Petition for dissolution of the IPE Corporation in the District Court of Tulsa County. Courtesy John Houchin.

now manufacturers are selling everything they can produce and as a result they are reluctant to advertise at a trade show."[6] Enright thought the difficulty at the 1979 show was two conflicting expositions, one at Aberdeen, Scotland, and a World Petroleum Congress in Bucharest. Enright concluded that he and fellow directors "are not the slightest bit interested in a traveling IPE" and that it would remain in Tulsa.[7]

Nevertheless, when the executive directors met on April 7, 1980, they approved a motion to close the exposition permanently and to dissolve its sponsoring corporation, blaming financial losses and competition from other shows. They recommended that the cash reserve of some $250,000 be distributed equally to Tulsa University, the University of Oklahoma, and Oklahoma State University in the form of petroleum related endowments. President John Houchin explained the directors' decision by noting that the shows of 1976 and 1979 had lost $254,958.95 and that a study by a market research firm indicated "that other expositions now satisfy the needs of exhibitors and their customers in the drilling and producing in-

When the IPE closed, its reserve funds were distributed to three universities. Receiving checks from IPE President John Houchin are (L to R) Dr. J. Paschal Twyman, president of the University of Tulsa; Charles Platt, Oklahoma State University Foundation; and Ron Burton, University of Oklahoma Foundation. Courtesy Leslie Brooks.

dustry. The IPE has a long history of which we can be justifiably proud," he concluded. "However, in light of changed conditions, a special committee recommended against holding future expositions and the executive directors reluctantly concurred."[8]

The IPE might well have continued into the 1980s in truncated, specialized—even regionalized—form. It would, of course, have had to find some way to break even, if not make a profit, but that would have meant abandoning the philosophy on which the IPE had been established. Perhaps W.K. Warren best summed up what transpired; when asked whether competing shows had "put the IPE out of business," he poignantly answered, "The oil industry put the IPE out of business."[9] Like a living organism, the oil industry had to adjust and then readjust to altered reality.

Some critics of the decision to close the IPE believe the great exposition was felled by boredom both among its volunteer industry leaders and its paid management, coupled with a proliferation of regional and specialized expositions.[10] These critics argue that there should have been a continuing infusion of vigorous, youthful leadership but that when those who built the IPE reached retirement age and could find no help forthcoming they preferred extinguishing the flame rather than watching it flicker and die.

Yet there were other reasons for the closing of the IPE, especially technological ones. When the IPE first opened in 1923, communication was slow, travel difficult, and oil field equipment huge and unwieldy. Suppliers and manufacturers could not conveniently come to buyers. But by the 1980s communications had become instantaneous, travel faster and easier, and the tools of the oil patch lighter and more portable; those which would not conveniently be transported could be shown easily by modern audio-visual technology. It had become possible for suppliers and manufacturers to come to—or very near to—oil men and save a great deal of money in the process.

The IPE in its fullness seemingly had become an anachronism, and those who closed it understood this. They thought it exemplified a brilliant epoch in the development of the oil industry worldwide, but it seemed to belong to an era that had passed into history. They voted to let it remain there.

And yet the oil industry is a curious thing, one where the only constant is dramatic and rapid change. Some who have looked at it like to compare it to a roller coaster, for it has high points filled with instant millionaires buying everything in sight and low periods when bankruptcy and overnight poverty are typical. Predicting these highs and lows is virtually impossible, as is foreseeing the future of any aspect of the oil industry. In 1920, for example, one agency of the federal government predicted that the world had only an eight-year supply of petroleum, while in 1978 the congress of the United States enacted lengthy legislation concerning the natural gas industry in the belief that the supply of this valuable commodity would be exhausted by 1990. Moreover, when the administration of Ronald Reagan pushed deregulation of the petroleum industry early in 1981, there were gloomy predictions of two-dollar-a-gallon gasoline. Deregulation brought a boom to the oil patch of almost mythic proportions in 1981-1982—leading to such success that by 1983 the world market was glutted with oil and natural gas, prices were falling at the pump, and manufacturers and suppliers again were hurting to make sales.

"What goes around will come around," once noted a sidewalk philosopher, perhaps with the oil industry in mind. By early 1985 there was talk of reviving the International Petroleum Exposition in Tulsa as a showcase for the newest, the brightest, the best in oil field technology and supplies. At the Tulsa State Fairgrounds the "Golden Driller" stands waiting, a silent sentinal to a brilliant memory—and just possibly a harbinger of the future.

NOTES

CHAPTER ONE

[1] Donald Barnum, "The International Petroleum Exposition, 1923-1940" (unpublished master's thesis, University of Tulsa, 1968), 26; hereafter cited as Barnum, *IPE*. This excellent study was of great value for background information on the first 11 meetings of the IPE. Interview, Leslie Brooks, June 4, 1982, Archives, Oklahoma Heritage Association (all interviews, unless otherwise noted, are in the OHA Archives).

[2] Barnum, "IPE," 40.
[3] *Ibid.*, 7.
[4] Kenny A. Franks, *The Oklahoma Petroleum Industry* (Oklahoma City: Oklahoma Heritage Association, 1980), 151.
[5] *Ibid.*
[6] *Ibid.*
[7] *Ibid.*
[8] *Tulsa Daily World*, October 7, 1923.
[9] Barnum, "IPE," 9.
[10] Interview, Leslie Brooks, June 4, 1982.
[11] Barnum, "IPE," 10; Fred S. Clinton, "The Beginnings of the International Petroleum Exposition and Congress," *Chronicles of Oklahoma*, XXVI (Winter 1948-1949), 481.
[12] Clinton, "Beginnings," 481.
[13] Barnum, "IPE," 11.
[14] Franks, *Oklahoma Petroleum Industry*, 151.
[15] *Ibid.*
[16] Barnum, "IPE," 12.
[17] *Ibid.*, 13.
[18] *Ibid.*, 14; Franks, *Oklahoma Petroleum Industry*, 152.
[19] Barnum, "IPE," 15.
[20] *Tulsa Tribune*, October 7, 1923.
[21] *Ibid.*, September 1, 1923.
[22] Franks, *Oklahoma Petroleum Industry*, 152.
[23] *Tulsa Daily World*, October 10, 11, 1923.
[24] *Tulsa Tribune*, September 27, 1923; Barnum, "IPE," 24.
[25] Franks, *Oklahoma Petroleum Industry*, 152, 153.
[26] *Tulsa Tribune*, September 1, 1923.
[27] She was attended by dutchesses Ann Kennedy, Lillian Randall, Katherine Gavin, Nell Cook, Rosaline Hollow, and Katherine Reisling.
[28] Barnum, "IPE," 28.
[29] "Petroleum Exposition Opens Monday," *Oil and Gas Journal*, October 4, 1923, 62.
[30] Interview, P.C. Lauinger, August 19, 1982.
[31] Barnum, "IPE," 31-32.
[32] *Ibid.*, 22.
[33] *Tulsa Daily World*, October 14, 1923.
[34] Barnum, "IPE," 38.
[35] *Ibid.*, 39.
[36] *Ibid.*, 48.
[37] *Ibid.*
[38] *Ibid.*, 50.
[39] Interview, Leslie Brooks, June 4, 1982.
[40] *Tulsa Tribune*, September 14, 1924.
[41] Carl Coke Rister, *Oil! Titan of the Southwest* (Norman: University of Oklahoma Press, 1949), 12-15; Barnum, "IPE," 50.

[42] *Tulsa Tribune*, October 1, 1924; Interview, Leslie Brooks, June 4, 1982.
[43] Barnum, "IPE," 45.
[44] *Tulsa Tribune*, September 14, 1924.
[45] *Ibid.*, October 1, 9, 1924.
[46] Barnum, "IPE," 54.
[47] *Tulsa Tribune*, October 3, 1924.
[48] Barnum, "IPE," 55-56.
[49] *Tulsa Tribune*, October 11, 1924.
[50] *Ibid.*, October 10, 1924.
[51] *Ibid.*, October 2, 1924; *Tulsa Daily World*, October 7, 1924.
[52] James McIntyre, "Oil Exposition Is an Immense Success," *Oil and Gas Journal*, October 2, 4. Items such as this one showing incomplete documentation were found in the scrapbooks compiled by various IPE officials over the years. These were put in the Special Collections Department of the Library at the University of Tulsa upon the dissolution of the International Petroleum Exposition Corporation and were loaned to the author during his research. Hereafter they will be cited as "IPE Scrapbook."
[53] *Tulsa Tribune*, October 5, 1924.
[54] *Tulsa Daily World*, October 7, 1924.
[55] Franks, *Oklahoma Petroleum Industry*, 156.
[56] *Ibid.*, 155.
[57] McIntyre, "Oil Immense Exhibit," *Oil and Gas Journal*, "IPE Scrapbook," 1924.
[58] *Tulsa Tribune*, October 5, 1924.
[59] *Tulsa Daily World*, October 3, 1924.
[60] *Ibid.*
[61] McIntyre, "Oil Immense Exhibit," 36.
[62] *Ibid.*, 37.
[63] Interview, Leslie Brooks, June 4, 1982.
[64] Franks, *Oklahoma Petroleum Industry*, 155.
[65] International Petroleum Industry, "History of the Oil Show," 4.
[66] James McIntyre, "Exposition Attracts Record Crowds," *Oil and Gas Journal*, October 9, 1924, 36.
[67] Franks, *Oklahoma Petroleum Industry*, 156.
[68] *Tulsa Tribune*, October 9, 1924.
[69] *Tulsa Daily World*, September 28, 1924.
[70] "Princesses Representing Oil Producing States at Exposition and Congress," *Oil and Gas Journal*, October 2, 1924, 40-41.
[71] *Tulsa Tribune*, September 27, 1924.
[72] Ralph T. Paker, "Petroleum Exposition Getting Ready," *Oil and Gas Journal*, September 11, 1924, 36.
[73] *Tulsa Tribune*, September 28, 1924.
[74] Barnum, "IPE," 64.
[75] *Ibid.*, 66.
[76] *Tulsa Tribune*, October 4, 1925.
[77] *Ibid.*, October 7, 1925; *Tulsa Daily World*, October 6, 1925.
[78] *Tulsa Daily World*, October 8, 1925.
[79] Barnum, "IPE," 73.
[80] *Tulsa Daily World*, October 4, 11, 1925.
[81] *Ibid.*, August 28, 1925.
[82] *Tulsa Tribune*, September 30, 1925.
[83] Barnum, "IPE," 68.

[84] *Tulsa Daily World*, October 4, 1925.
[85] *Ibid.*, October 9, 1925.
[86] *Tulsa Tribune*, October 2, 1925.
[87] *Tulsa Daily World*, October 4, 1925.
[88] Interview, Leslie Brooks, June 18, 1982.
[89] *Tulsa Daily World*, October 2, 1925.
[90] *Ibid.*, October 3, 4, 9, 1925.
[91] *Ibid.*, October 4, 1925.
[92] Barnum, "IPE," 82.
[93] *Tulsa Daily World*, October 11, 1925.
[94] Barnum, "IPE," 84; "History of the Oil Show," 4.
[95] Barnum, "IPE," 84.
[96] *Ibid.*
[97] "History of the Oil Show," 5.
[98] Barnum, "IPE," 86.
[99] *Ibid.*, 87.
[100] *Tulsa Daily World*, September 25, 1927.
[101] Interview, Leslie Brooks, June 4, 1982.
[102] *Tulsa Tribune*, September 25, 1927; *Tulsa Daily World*, September 30, 1927.
[103] *Tulsa Daily World*, September 25, 1927.
[104] *Tulsa Tribune*, September 25, 1927.
[105] *Ibid.*
[106] *Tulsa Daily World*, September 26, 1927.
[107] *Tulsa Tribune*, September 25, 1927.
[108] *Ibid.*, September 28, 1927.
[109] *Tulsa Daily World*, September 25, 1927.
[110] *Tulsa Tribune*, September 27, 1927.
[111] *Ibid.*, September 30, 1927.
[112] *Ibid.*
[113] *Ibid.*
[114] *Ibid.*, October 2, 1927.
[115] *Tulsa Daily World*, May 14, 1953.
[116] *Ibid.*, September 25, 1927; Barnum, "IPE," 97.
[117] *Tulsa Tribune*, October 2, 1927.
[118] "History of the Oil Show," 5.
[119] Barnum, "IPE," 99.
[120] *Tulsa Daily World*, October 3, 1927.
[121] *Ibid.*, October 21, 1928.
[122] Barnum, "IPE," 106.
[123] *Ibid.*, 102.
[124] *Ibid.*, 100.
[125] *Tulsa Daily World*, October 21, 1928.
[126] *Ibid.*
[127] Barnum, "IPE," 105.
[128] *Tulsa Tribune*, October 25, 1928.
[129] *Ibid.*, October 23, 1928; *Tulsa Daily World*, October 24, 1928.
[130] *Tulsa Daily World*, October 24, 1928.
[131] *Ibid.*, May 14, 1953.
[132] "Our Fifth IPE," *Tulsa Spirit*, September 24 (1928 IPE Scrapbook).
[133] *Tulsa Tribune*, October 28, 1928.
[134] *Ibid.*, October 17, 21, 1928.
[135] "Fifth IPE and Congress," *This Week in Tulsa*, October 20, 1928 (1928 IPE Scrapbook).
[136] *Tulsa Tribune*, October 30, 1928.
[137] *Ibid.*
[138] Ben Harrison, "IPE Has Colorful History," *Tulsa*, XL (May 1966), 16.
[139] Barnum, "IPE," 109.
[140] *Ibid.*
[141] *Ibid.*
[142] *Ibid.*, 108-110.
[143] *Tulsa Tribune*, October 5, 6, 1929.
[144] *Ibid.*, October 11, 1929.
[145] Barnum, "IPE," 119.
[146] "Men Behind Tulsa's Big Oil Show," *Oil Weekly*, May 16, 1938 (1938 IPE Scrapbook).
[147] *Tulsa Tribune*, October 2, 1929.
[148] *Tulsa Daily World*, October 2, 5, 1929.
[149] *Ibid.*, October 6, 1929.
[150] *Ibid.*
[151] Rister, *Oil!*, 254.
[152] *Tulsa Tribune*, October 8, 13, 1929.
[153] *Ibid.*, October 7, 1929.
[154] *Ibid.*, October 6, 1929; *Tulsa Daily World*, October 6, 1929.
[155] *Tulsa Tribune*, October 10, 1929.
[156] *Ibid.*, October 11, 1929.
[157] *Tulsa Daily World*, October 6, 1929.
[158] Barnum, "IPE," 120.
[159] *Ibid.*, 121.
[160] *Tulsa Tribune*, October 5, 1930.
[161] *Ibid.*
[162] "History of the Oil Show," 7.
[163] *Ibid.*
[164] *Tulsa Tribune*, October 5, 1930.
[165] *Ibid.*, May 13, 1934.
[166] *Ibid.*, October 5, 1930.
[167] Barnum, "IPE," 122.
[168] *Tulsa Tribune*, October 5, 1930.
[169] *Ibid.*, October 8, 1930.
[170] *Ibid.*, October 10, 1930.
[171] *Ibid.*, October 9, 1930.
[172] *Ibid.*
[173] Barnum, "IPE," 128.
[174] *Tulsa Tribune*, October 9, 9, 11, 1930.
[175] Barnum, "IPE," 128.
[176] *Ibid.*

CHAPTER TWO

[1] "Eighth International Petroleum Exposition," *National Petroleum News*, May 9, 1934, 8 (1934 IPE Scrapbook).
[2] "Exhibitors Take Larger Space for Oil Show," *Petroleum News*, January 10, 1934, 12 (1934 IPE Scrapbook).
[3] "International Petroleum Exposition Expected To Be Large," *Petroleum Engineer*, December 1933, 17 (1934 IPE Scrapbook).
[4] "Exhibitors Take Larger Space for Oil Show," 12.
[5] *Ibid.*
[6] "President Roosevelt To Proclaim Oil Exposition," *Oil and Gas Journal*, February 8, 1934, 39.
[7] "Many Methods Being Used To Assure Tulsa Show Attendance," *Petroleum Engineer*, March 1934, 20 (1934 IPE Scrapbook).
[8] "Exposition Committee To Tell State of Oil Show," *Tulsa Spirit*, March 28, 1934, 8 (1934 IPE Scrapbook).
[9] "Associations Meet During Oil Show," *Petroleum News*, January 24, 1934, 10 (1934 IPE Scrapbook).
[10] "World's Fair of the Oil Industry," *Oil and Gas Journal*, May 10, 1934, 31.
[11] Unidentified clipping, 1934 IPE Scrapbook.
[12] *Tulsa Daily World*, May 13, 1934.

[13] *Ibid.*, May 15, 1934.
[14] Unidentified clipping, 1934 IPE Scrapbook.
[15] *Tulsa Daily World*, May 13, 1934.
[16] *Tulsa Tribune*, May 13, 1934.
[17] Unidentified clipping, 1934 IPE Scrapbook.
[18] *Tulsa Daily World*, May 13, 1934.
[19] Barnum, "IPE," 143.
[20] "World's Greatest Petroleum Exposition," *Manufacturer's Record*, May 1934 (1934 IPE Scrapbook).
[21] *Ibid.*
[22] "Let's Talk It Over," *Super Service Station*, March 1934, 5 (1934 IPE Scrapbook).
[23] *Ibid.*
[24] "World's Greatest Petroleum Exposition."
[25] "World's Fair of the Oil Industry to Be Held At Tulsa," *Skelly News*, January-February-March 1934 (1934 IPE Scrapbook).
[26] *Ibid.*
[27] Unidentified clipping, 1934 IPE Scrapbook.
[28] *Ibid.*
[29] "See Catalog in Motion at Tulsa Oil Exposition," *Oil and Gas Journal*, February 8, 1934, 50.
[30] *Ibid.*
[31] "Committee for First-Aid Contest To Be Held at Tulsa Exposition," *Ibid.*, February 15, 1934.
[32] *Ibid.*
[33] *Ibid.*
[34] *Tulsa Daily World*, May 15, 1934.
[35] *Ibid.*
[36] *Ibid.*
[37] Barnum, "IPE," 148.
[38] "Oil Capitol Horse Show," *The Southwestern Horseman*, April 1934, 23 (1934 IPE Scrapbook).
[39] Barnum, "IPE," 149.
[40] *Ibid.*
[41] *Tulsa Tribune*, May 17, 1934.
[42] "Tulsa Exposition Greatest Yet, According to Exhibitors," unidentified clipping, 1934 IPE Scrapbook.
[43] *Elk City (Oklahoma) News*, May 28, 1934.
[44] "World's Fair of the Oil Industry To Be Held at Tulsa."
[45] "International Petroleum Exposition To Open Its Gates Saturday," *Oil and Gas Journal*, May 14, 1936, 109.
[46] *Ibid.*
[47] *Ibid.*
[48] *Ibid.*
[49] Barnum, "IPE," 131.
[50] "Ninth International Petroleum Exposition To Open May 16," *Oil and Gas Journal*, May 7, 1936, 114.
[51] *Ibid.*
[52] *Tulsa Tribune*, May 16, 1936.
[53] Barnum, "IPE," 135.
[54] "New, Greater Display Announced for International Petroleum Exposition," *Oil and Gas Journal*, April 16, 1936, 60.
[55] *Ibid.*
[56] *Ibid.*
[57] *Tulsa Tribune*, May 20, 1936.
[58] "History of the Oil Show," 9.
[59] "New, Greater Display Announced," 60.
[60] "Program Directory" for 1936 IPE, 2, 1936 IPE Scrapbook.
[61] Interview, William J. Sherry, July 9, 1982.
[62] "Old Timers Are Much Interested in Awards at Petroleum Exposition," *Oil and Gas Journal*, April 23, 1936, 20.
[63] *Ibid.*, 19.

[64] "Old Timers of Industry and of Tulsa Are Honored at Special Ceremony," *Oil and Gas Journal*, May 28, 1936, 19.
[65] *Ibid.*
[66] Unidentified clipping, 1936 IPE Scrapbook.
[67] *Ibid.*
[68] *Ibid.*
[69] *Ibid.*
[70] *Tulsa Tribune*, May 19, 1936.
[71] *Ibid.*, May 13, 1936.
[72] *Ibid.*, May 14, 1936.
[73] Barnum, "IPE," 151.
[74] *Tulsa Tribune*, May 21, 1936.
[75] *Tulsa Daily World*, May 24, 1940.
[76] "Fifty Percent Larger International Petroleum Exposition at Tulsa, May 14-21," *The Oil Expositor*, May 20, 1938, 1.
[77] *Tulsa Daily World*, May 18, 1937; May 1, 1938.
[78] "Oil Exposition Ready for Opening Saturday," unidentified clipping, 1938 IPE Scrapbook.
[79] *Tulsa Tribune*, May 15, 1938.
[80] *Ibid.*, May 10, 1938.
[81] *Ibid.*
[82] *Ibid.*, May 17, 1938.
[83] Unidentified clipping, 1938 IPE Scrapbook.
[84] *Ibid.*
[85] *Tulsa Tribune*, May 15, 1938.
[86] *Tulsa Daily World*, May 15, 1938.
[87] *Ibid.*
[88] *Ibid.*
[89] *Ibid.*
[90] *Ibid.*
[91] *Tulsa Tribune*, May 6, 1938.
[92] *Ibid.*, May 13, 1938.
[93] *Ibid.*, May 16, 1938.
[94] Unidentified clipping, 1938 IPE Scrapbook.
[95] *Tulsa Daily World*, May 15, 1938.
[96] Unidentified clipping, 1938 IPE Scrapbook.
[97] *Tulsa Daily World*, May 15, 20, 1938.
[98] *Tulsa Tribune*, May 12, 1938.
[99] "Prepare for Large Attendance from All Nations at the Tulsa Oil Show," *Tulsa Oil Expositor*," April 1938, 1938 IPE Scrapbook.
[100] *Ibid.*
[101] *Tulsa Tribune*, May 18, 1940.
[102] *Tulsa Daily World*, May 19, 1940.
[103] *Ibid.*
[104] "Prepare for Large Attendance from All Nations at the Tulsa Oil Show," 1938 IPE Scrapbook.
[105] *Ibid.*
[106] *Ibid.*
[107] *Ibid.*
[108] Barnum, "IPE," 146.
[109] "Prepare for Large Attendance."
[110] *Tulsa Tribune*, May 15, 1938.
[111] *Tulsa Daily World*, May 16, 1938.
[112] *Tulsa Tribune*, May 20, 1938.
[113] *Ibid.*, May 12, 1938.
[114] "Prepare for Large Attendance."
[115] *Tulsa Daily World*, May 6, 1938.
[116] Unidentified clipping, 1938 IPE Scrapbook.
[117] *Tulsa Tribune*, May 15, 1938.
[118] "Information Souvenir of the World's Gala Oil Show," *Tulsa Guide*, May 13-19, 1938 IPE Scrapbook.
[119] Unidentified clipping, 1938 IPE Scrapbook.
[120] *Tulsa Tribune*, May 12, 1938.
[121] *Ibid.*, May 18, 1940.

¹²²Baron Creager, "Expo Visitor Predicts Collapse of Germany from Shortage of Oil," unidentified clipping, 1940 IPE Scrapbook.
¹²³"Public Side of Exposition Emphasized Much More," *Independent Monthly*, unidentified clipping, 1940 IPE Scrapbook.
¹²⁴*Ibid.*
¹²⁵*Tulsa Tribune*, December 31, 1939.
¹²⁶*Tulsa Daily World*, September 10, 1939.
¹²⁷*Ibid.*, April 10, 1940.
¹²⁸*Ibid.*
¹²⁹*Ibid.*, May 19, 1940.
¹³⁰*Ibid.*
¹³¹*Ibid.*
¹³²*Ibid.*
¹³³Barnum, "IPE," 140.
¹³⁴*Tulsa Daily World*, May 19, 1940.
¹³⁵*Ibid.*
¹³⁶*Tulsa Tribune*, May 18, 1940.
¹³⁷"Expo Will Unveil Newest in Equipment," *National Petroleum News*, April 1940, clipping, 1940 IPE Scrapbook.
¹³⁸*Petroleum Equipment*, January 1940, clipping, 1940 IPE Scrapbook.
¹³⁹"May Open Branch Offices in Tulsa," clipping, 1940 IPE Scrapbook.
¹⁴⁰"Tulsa Exposition Standing Committees," *Oil Marketer*, January 15, 1940, 1940 IPE Scrapbook.
¹⁴¹"Big Model Factory Promised Oil Expo," unidentified clipping, 1940 IPE Scrapbook.
¹⁴²"U.S. Steel Will Display Products at Expo," unidentified clipping, 1940 IPE Scrapbook.
¹⁴³"Inventor of Glass-Topped Gasoline Pump First to Sign," unidentified clipping, 1940 IPE Scrapbook.
¹⁴⁴"Expo Will Unveil Newest in Equipment."
¹⁴⁵*Tulsa Tribune*, December 10, 1939.
¹⁴⁶San Antonio *Texas News Globe*, December 3, 1939.
¹⁴⁷Barnum, "IPE," 147.
¹⁴⁸"Shower of Flags at Oil Show," unidentified clipping, 1940 IPE Scrapbook.
¹⁴⁹*Ibid.*
¹⁵⁰Interview, Leslie Brooks, June 18, 1982.
¹⁵¹Creager, "Expo Visitor Predicts Collapse of Germany."
¹⁵²*Tulsa Tribune*, May 19, 1940.
¹⁵³Unidentified clipping, 1940 IPE Scrapbook.
¹⁵⁴Mather Eakes, Jr., "Tulsa Invention Will Bring Great Oil Well Saving," unidentified clipping, 1940 IPE Scrapbook.
¹⁵⁵*Ibid.*
¹⁵⁶*Tulsa Tribune*, May 19, 1940.
¹⁵⁷*Ibid.*
¹⁵⁸*Ibid.*
¹⁵⁹*Ibid.*
¹⁶⁰*Ibid.*
¹⁶¹*Ibid.*
¹⁶²*Ibid.*, May 18, 1940.
¹⁶³*Ibid.*
¹⁶⁴*Ibid.*
¹⁶⁵*Ibid.*
¹⁶⁶"IPE Plans for Visiting Writers," unidentified clipping, 1940 IPE Scrapbook.
¹⁶⁷"Honors Pioneers at Exposition," unidentified clipping, 1940 IPE Scrapbook.
¹⁶⁸*Tulsa Daily World*, May 25, 1940.
¹⁶⁹*Ibid.*
¹⁷⁰"Sixty Students to Visit Exposition," unidentified clipping, 1940 IPE Scrapbook.
¹⁷¹"Tulsa Chapter of NOMADS to Organize January 8," *Oil Weekly*, 1940 IPE Scrapbook.
¹⁷²*Tulsa Daily World*, February 26, 1940.
¹⁷³*Ibid.*, July 30, 1939.
¹⁷⁴*Ibid.*, May 22, 1940.

CHAPTER THREE

¹*Tulsa Daily World*, January 11, 18, 1948.
²*Ibid.*, November 16, 1947.
³*Ibid.*, April 5, July 13, 1947.
⁴*Ibid.*, July 13, 1947.
⁵*Ibid.*, July 19, 1947.
⁶Unidentified clipping, 1948 IPE Scrapbook.
⁷"The Petroleum Industry and DuPont," *DuPont Magazine*, April 1948, clipping, 1948 IPE Scrapbook.
⁸*Tulsa Daily World*, May 4, 1948.
⁹*Tulsa Tribune*, July 1, 1947.
¹⁰*Ibid.*
¹¹*Ibid.*, March 9, 1948.
¹²Tulsa Daily World, May 14, 1948.
¹³*Ibid.*
¹⁴*Ibid.*
¹⁵*Tulsa Tribune*, May 5, 1948.
¹⁶*Tulsa Daily World*, May 14, 1948.
¹⁷"IPE Officials Announce Open Forum Oil Congress," *Tulsa Spirit*, February 24, 1948, copy in 1948 IPE Scrapbook.
¹⁸*Ibid.*; *Tulsa Daily World*, February 8, 1948.
¹⁹"Plans Outlined for International Oil Congress at Exposition," *Oil and Gas Journal*, April 22, 1948, copy in 1948 IPE Scrapbook.
²⁰*Ibid.*
²¹*Ibid.*
²²Ted Armstrong, "Varying Oil Policies in the Americas Defended by IPE Congress Speakers," *Oil and Gas Journal*, May 27, 1948, copy in 1948 IPE Scrapbook.
²³Unidentified clipping, 1948 IPE Scrapbook.
²⁴"NOMADS Active in Planning Their Part in Petroleum Exposition," *Petroleum Equipment*, April-May-June 1948, copy in 1948 IPE Scrapbook.
²⁵*Ibid.*
²⁶*Tulsa Daily World*, April 26, 1948.
²⁷*Ibid.*, September 10, 1947.
²⁸*Ibid.*, December 8, 1947.
²⁹*Ibid.*, February 1, 1948.
³⁰*Ibid.*
³¹"300,000 Visitors Throng IPE," *Oil and Gas Journal*, May 27, 1948, copy in 1948 IPE Scrapbook.
³²*Ibid.*
³³*Tulsa Daily World*, April 11, 1948.
³⁴"300,000 Visitors Throng IPE."
³⁵*Tulsa Tribune*, April 28, 1948.
³⁶Unidentified clipping, 1948 IPE Scrapbook.
³⁷*Ibid.*
³⁸*Ibid.*
³⁹*Ibid.*
⁴⁰*Tulsa Tribune*, May 14, 1948.
⁴¹*Tulsa Daily World*, November 9, 1947.
⁴²*Ibid.*, May 13, 1948.
⁴³*Ibid.*, May 18, 1948.

44"Hall of Science Portrays Industry's Progress," *Oil and Gas Journal*, May 1948, copy in 1948 IPE Scrapbook.
45*Tulsa Daily World*, May 23, 1948.
46*Ibid.*
47*Ibid.*, June 20, 1951.
48*Ibid.*, June 21, 1951.
49*Tulsa Tribune*, June 9, 1952.
50"IPE Housing Bureau Launches Campaign for Early Reservations for Oil Show," *Tulsa Spirit*, December 25, 1952, copy in 1953 IPE Scrapbook.
51*Ibid.*
52*Tulsa Daily World*, June 6, 1948.
53"Petroleum Exposition Expanding Facilities for Its 1953 Show," *Texas Oil Journal*, September 1952, copy in 1953 IPE Scrapbook.
54*Tulsa Daily World*, August 15, 1952.
55"Think Deep—Think Big—Think July," *Drilling: The Wellsite Publication*, May 1976, copy in 1953 IPE Scrapbook.
56Henry D. Ralph, "Huge Show Under Way," *Oil and Gas Journal*, May 18, 1953, copy in 1953 IPE Scrapbook.
57"Huge Show Under Way."
58*Ibid.*
59*Ibid.*
60*Ibid.*
61*Tulsa Daily World*, July 27, 1952.
62"Huge Show Under Way."
63*Ibid.*
64*Ibid.*
65*Ibid.*
66*Ibid.*
67*Ibid.*
68"Clinic for Oil Men," *Oil and Gas Journal*, May 11, 1953, copy in 1953 IPE Scrapbook.
69*Tulsa Daily World*, November 23, 1952.
70Philip C. Ingalls, "Education in Oil," *Oil and Gas Journal*, May 18, 1953, copy in 1953 IPE Scrapbook.
71*Ibid.*
72*Ibid.*
73*Ibid.*
74*Ibid.*
75*Ibid.*
76*Ibid.*
77"Oil Films Shown," *Oil and Gas Journal*, May 18, 1953, copy in 1953 IPE Scrapbook.
78"Special Events Slated," *Ibid.*
79*Ibid.*
80*Tulsa Daily World*, June 29, 1952.
81*Ibid.*
82*Ibid.*, November 9, 1952.
83"Huge Show Under Way."
84*Ibid.*
85*Tulsa Daily World*, December 21, 1952.
86*Ibid.*
87IPE, "Awardees of 'Old Timers.'"
88*Tulsa Daily World*, August 29, 1957.

89Ruth Sheldon Knowles, *The Greatest Gamblers* (New York: McGraw-Hill, 1959), v.
90*Tulsa Daily World*, August 29, 1957.
91Interview, W.K. Warren, May 6, 1983.
92*Tulsa Tribune*, May 2, 1957.
93*Ibid.*, November 1, 1957.
94"W.K. Warren and 'Kep' Keplinger Tell What Tulsa Is Doing," *Tulsa Gasser*, April 25, 1959, copy in 1959 IPE Scrapbook.
95*Ibid.*
96"Tulsa's Debt to Oil," *Greater Tulsa*, May 7, 1959, copy in 1959 IPE Scrapbook.
97*Tulsa Daily World*, June 15, 1958.
98*Ibid.*, May 13, 1959.
99"IPE Spending for Largest Oil Show 'Big Business' for Tulsa," *Tulsa Spirit*, March 18, 1959, copy in 1959 IPE Scrapbook.
100*Tulsa Daily World*, March 9, 1958.
101*Tulsa Tribune*, May 14, 1959.
102*Ibid.*
103*Ibid.*
104*Ibid.*
105*Ibid.*
106*Ibid.*
107*Ibid.*
108*Ibid.*
109Unidentified clipping, 1959 IPE Scrapbook.
110*Ibid.*
111*Tulsa Tribune*, May 18, 1959.
112*Ibid.*
113*Ibid.*
114*Ibid.*
115*Ibid.*
116*Tulsa Daily World*, May 16, 1959.
117*Ibid.*, May 14, 1959.
118*Tulsa Tribune*, May 8, 1959.
119*Tulsa Daily World*, January 24, April 17, May 12, 1959.
120*Ibid.*, May 18, 1959.
121*Ibid.*, May 22, 1959; unidentified clipping, 1959 IPE Scrapbook.
122*Tulsa Daily World*, May 14, 1959.
123*Tulsa Tribune*, April 9, 1959.
124*Ibid.*
125*Ibid.*
126*Ibid.*
127*Ibid.*
128*Ibid.*
129*Ibid.*, April 23, 1959.
130"Mrs. Bridgeman 'Oil Woman of the Year,'" *Oil Daily*, April 22, 1959, copy in 1959 IPE Scrapbook.
131*Tulsa Tribune*, May 13, 1959.
132*Ibid.*, April 6, 1959.
133"Thirty Thousand Oil Men Expected," *Petroleum Week*, January 23, 1959, copy in 1959 IPE Scrapbook.

CHAPTER FOUR

1Interview, W.K. Warren, May 6, 1983.
2*Tulsa Tribune*, August 16, 1963.
3*Ibid.*
4*Ibid.*
5*Ibid.*
6*Ibid.*, October 20, 1963.
7*Ibid.*

8"Lease Unites Fair, IPE Ground," unidentified clipping, 1966 IPE Scrapbook.
9*Ibid.*
10*Tulsa Daily World*, November 8, December 2, 1963; *Tulsa Tribune*, December 4, 1963.
11*Tulsa Tribune*, December 4, 1963.
12*Tulsa Daily World*, April 4, 1966.

[13] Ibid.
[14] Ibid., April 11, 1966.
[15] "IPE Adopts 'Golden Driller' as Show Symbol," *Greater Tulsa*, December 2, 1965, copy in 1966 IPE Scrapbook.
[16] *Tulsa Tribune*, June 10, 1964.
[17] "IPE Exhibitor Space Cost 10 Percent Below 1959," *Oil Daily*, July 8, 1964, copy in 1966 IPE Scrapbook.
[18] *Tulsa Tribune*, March 1, 1966.
[19] *Tulsa Daily World*, June 11, 1964.
[20] *Tulsa Tribune*, January 13, 1965.
[21] *Tulsa Daily World*, March 16, 1965.
[22] *Tulsa Tribune*, May 12, 1966.
[23] Ibid., March 12, 1966.
[24] Ibid., May 12, 1966.
[25] *Tulsa Daily World*, May 13, 1966.
[26] *Tulsa Tribune*, May 12, 1966.
[27] *Tulsa Daily World*, May 13, 1966.
[28] Ibid.
[29] Ibid.
[30] Ibid.
[31] *Tulsa Daily World* May 15, 1966.
[32] Ibid.
[33] *Tulsa Tribune*, May 13, 1966.
[34] Ibid.
[35] *Tulsa Daily World*, May 15, 1966.
[36] Ibid.
[37] Ibid.
[38] Ibid.
[39] Ibid.
[40] *Tulsa Tribune*, May 19, 1966.
[41] *Tulsa Daily World*, May 30, 1966.
[42] Ibid.
[43] *Tulsa Tribune*, June 14, 1966.
[44] *Tulsa Daily World*, May 1, 1966.
[45] *Tulsa Tribune*, April 29, 1966.
[46] *Tulsa Daily World*, December 16, 1965.
[47] Ibid., May 22, 1966.
[48] Ibid.
[49] Ibid.
[50] *Tulsa Tribune*, May 26, 1966.
[51] "International Petroleum Exposition makes Special Effort To Attract Independents to 1971 Show at Tulsa, May 15, 23," *Mid-America Oil and Gas Reporter* (February 1971), 6.
[52] Ibid.
[53] Ibid.
[54] *Tulsa Daily World*, December 27, 1970.
[55] "Yost Re-elected IPE President," *Drilling-DCW* (March 9, 1971), 9.
[56] *Tulsa Tribune*, January 22, 1971.
[57] "IPE Exhibitors Sign Early for 1971 Show," *Mid-America Oil and Gas Reporter* (December 1969), copy in 1971 IPE Scrapbook.
[58] *Tulsa Daily World*, December 27, 1970.
[59] Ibid.
[60] Ibid., May 16, 1971.
[61] Ibid.
[62] Ibid., December 27, 1970.
[63] Ibid.
[64] Ibid.
[65] "All Phases of Petroleum Featured at 1971 Exposition," *The Drill Bit*, copy in 1971 IPE Scrapbook.
[66] *Tulsa Daily World*, January 22, 1971.
[67] Ibid.
[68] Ibid.
[69] Ibid., May 21, 1971.
[70] Ibid., May 17, 1971.
[71] Ibid., May 21, 1971.
[72] Ibid.
[73] *Tulsa Tribune*, May 21, 1971.
[74] Ibid.
[75] *Tulsa Daily World*, December 27, 1970.
[76] "Fourteen British Firms To Exhibit at 1971 Tulsa Oil Show," *Tulsa Engineer* (September 1970), 9.
[77] Ibid.
[78] *Tulsa Daily World*, December 17, 1970.
[79] Ibid.
[80] "Activities Chairman Announced IPE 'First Lady of Petroleum' Contest," *Independent Petroleum Monthly*, February 1971, copy in 1971 IPE Scrapbook.
[81] *Tulsa Daily World*, May 20, 1971.
[82] Ibid.
[83] IPE, "Grand Old Man Awardees."
[84] *Tulsa Daily World*, May 16, 1971.
[85] Ibid.
[86] *Tulsa Tribune*, May 24, 1971.
[87] Ibid.
[88] *Tulsa Daily World*, May 21, 1971.
[89] "Oil Capitol Hosts IPE," unidentified clipping, 1976 IPE Scrapbook.
[90] Ibid.
[91] IPE, "Energy '76 Floor Plan," 1976 IPE Scrapbook.
[92] *Tulsa Daily World, May 16, 18, 1976; Tulsa Tribune*, May 17, 1976.
[93] *Tulsa Tribune*, May 13, 1976.
[94] Ibid.
[95] Ibid.
[96] Ibid.
[97] Ibid.
[98] Ibid.
[99] *Tulsa Daily World*, May 16, 1976.
[100] "IPE: Where the Accent is Onshore," *Drilling-DCW* (april 1976), 21; copy in 1976 IPE Scrapbook.
[101] *Tulsa Daily World*, February 23, 1976.
[102] IPE, "NOMADS Energy Technical Symposium Agenda," 1976 IPE Scrapbook.
[103] Ibid.
[104] Ibid.
[105] Ibid.
[106] Ibid.
[107] Ibid.
[108] "IPE: Where the Accent is Onshore."
[109] IPE, "News Release," 1976 IPE Scrapbook.
[110] Ibid.
[111] Ibid.
[112] Ibid.
[113] Ibid.
[114] *Tulsa Daily World*, November 11, 1975.
[115] *Tulsa Tribune*, November 20, 1975.
[116] *Tulsa Daily World*, February 16, 1976.
[117] Ibid.
[118] *Tulsa Tribune*, January 19, 1976.
[119] Ibid.
[120] Ibid.
[121] Ibid.
[122] Ibid., April 22, 1976.
[123] Telegram, J. Paul Getty to William J. Sherry, 1976 IPE Scrapbook.
[124] *Tulsa Tribune*, May 22, 1976.
[125] "Petroleum Exposition Was A 'Gusher,'" *Southwest Industrial News* (June 9, 1976), 1, copy in 1976 IPE Scrapbook.

[126] Summary of survey results, in private papers of Leslie Brooks.
[127] IPE, "Fact Sheet on the 1979 IPE," 1979 IPE Scrapbook.
[128] *Tulsa Tribune*, January 31, 1978.
[129] IPE, "Invitation—International Onshore Petroleum Exposition and Congress," 1979 IPE Scrapbook.
[130] *Tulsa Daily World*, April 24, 1979.
[131] *Ibid.*
[132] *Ibid.*
[133] IPE, "Exhibitors Approve a Four-Day Show," *IPE Energy '79 Newsletter*, 1979 IPE Scrapbook.
[134] IPE, "Three Publications Plan Special Issues," *Ibid.*
[135] "Multi-Lanugage IPE Brochures Printed," *Ibid.*
[136] *Tulsa Daily World*, August 22, 1979.
[137] Interview, James O. Kemm, September 21, 1982.
[138] *Tulsa Daily World*, August 22, 1979.
[139] *Tulsa Tribune*, September 10, 1979.
[140] "Invitation—International Onshore Petroleum Exposition and Congress."
[141] "Plan Enhanced Recovery Hall of Science for IPE," *IPE Energy '79 Newsletter*, 1979 IPE Scrapbook.
[142] *Ibid.*
[143] Interview, James O. Kemm, September 21, 1982.
[144] *Tulsa Daily World*, September 14, 1979.
[145] "Simon To Speak at IPE Seminar," unidentified clipping, 1979 IPE Scrapbook.
[146] "News Release: EST Plans Special Seminar on Productivity at Tulsa Oil Show," 1979 IPE Scrapbook.
[147] *Ibid.*
[148] "Moderators Chosen for IPE Symposium," *IPE Energy '79 Newsletter*, 1979 IPE Scrapbook.
[149] "Here's a Preview of Some Exhibits Being Displayed at '79 International Petroleum Exposition," *The Oil Daily* (July 9, 1979), 2; copy in 1979 IPE Scrapbook.
[150] *Tulsa Daily World*, September 7, 1979.
[151] *Ibid.*
[152] *Ibid.*

EPILOGUE

[1] *Tulsa Tribune*, September 14, 1979.
[2] "Larger Crowds Please IPE Officials, Exhibitors," unidentified clipping, 1979 IPE Scrapbook.
[3] *Denver Post*, November 17, 1979.
[4] *Ibid.*
[5] *Tulsa Tribune*, November 16, 1979.
[6] *Ibid.*
[7] *Ibid.*
[8] *Tulsa Daily World*, April 16, 1979.
[9] Interview, W.K. Warren, May 6, 1983.
[10] Interview, Herbert C. Fries and Otha Grimes, June 9, 1983.

BIBLIOGRAPHY

THE IPE SCRAPBOOKS

Much of the material used in this book was taken from the scrapbooks kept through the years by the managers and staff employees of the IPE and by Leslie Brooks, who assumed responsibility for IPE publicity in 1934. The scrapbooks contain newspaper clippings, articles, IPE news releases and newsletters, correspondence by IPE officials, telegrams, and IPE promotional brochures. When the IPE closed in 1979, these scrapbooks were deposited in the Special Collections Department of the Library of the University of Tulsa. No research on the IPE would be complete without reference to these scrapbooks.

UNPUBLISHED MATERIALS

Barnum, Donald meyer, "International Petroleum Exposition: 1923-1940," Master of Arts Thesis, University of Tulsa, 1968.

Leslie Brooks, private papers.

William J. Sherry, private papers.

INTERVIEWS

William Bell, August 18, 1982.
Leslie Brooks, June 4, 1982.
Herbert C. Fries, June 9, 1983.
Otha Grimes, June 9, 1983.
John Houchin, July 19, 1982.
James O. Kemm, September 21, 1982.
P.C. Lauinger, August 19, 1982.
William J. Sherry, July 9, 1982.
W.K. Warren, May 6, 1983.
Randolph Yost, August 10, 1982.

(The transcripts of these interviews are in the Archives, Oklahoma Heritage Association).

NEWSPAPERS

Denver Post
Elk City News (Elk City, Oklahoma)
Texas News Globe (San Antonio)
Tulsa Daily World
Tulsa Tribune

PUBLISHED MATERIAL

"Associations Meet During Show," *Petroleum News* (January 24, 1934), 10.

Clinton, Fred S. "The Beginnings of the International Petroleum Exposition and Congress," *Chronicles of Oklahoma* (Winter 1948-1949), 479-388.

"Eighth International Petroleum Exposition," *National Petroleum News* (May 9, 1934), 8.

"Exhibitors Take Larger Space for Oil Show," *Petroleum News8M (January 10, 1934)*, 12.

"*Fifty Percent Larger International Petroleum Exposition at Tulsa, May 14 21*," *The Oil Expositor* (May 20, 1938), 1-5.

"Fourteen British Firms To Exhibit at 1971 Tulsa Oil Show," *Tulsa Engineer* (September 1970), 9-16.

Faulk, Odie B., James H. Thomas, and Carl N. Tyson. *The Gentleman: The Life of Joseph A. LaFortune*. Oklahoma City: Oklahoma Heritage Association, 1979.

Franks, Kenny A. *The Oklahoma Petroleum Industry*. Oklahoma City: Oklahoma Heritage Association, 1980.

Harrison, Ben. "IPE Has Colorful History," *Tulsa* (May 1966), 16-20.

Knowles, Ruth Sheldon. *The Greatest Gamblers*. New York: McGraw-Hill, 1959.

Oil and Gas Journal. See issues for each year of IPE.

Rister, Carl Coke. *Oil! Titan of the Southwest*. Norman: University of Oklahoma Press, 1959.

"See Catalog in Motion at Tulsa Oil Exposition," *Oil and Gas Journal* (February 8, 1934), 50.

"World's Fair of the Oil Industry," *Oil and Gas Journal* (May 10, 1934), 31-35.

"Yost Re-elected IPE President," *Drilling-DCW* (March 9, 1971), 9-10.

INDEX

Aaron Machinery Company, 83
Abernathy, Jack H., 97
Adair, Paul, 95
Adams, Ernestine, 80
Adams, W. Thomas, 103
Agrico Chemical Co., 105, 109
Akdar Shriners, Tulsa, 22
Alaska, 97, 99, 103
Albert, E.R., Jr., 107
Alexander, Louis, 108
Allen, Henry J., 17
Allied Helicopter Service, 86
Allis-Chalmers Co., 52-53
Alman, G.D., Jr., 97
Alyeska Corp., 103
Amerada Petroleum Corp., 53, 60, 94
American Airlines, 52
American Association of Petroleum Geologists, 38
American Car and Foundary Co., 19
American Chemical Society, 73
American Gas Association, 33, 38
American Institute of Mining and Metallurgical Engineers, 46
American Petroleum Institute, 17, 25, 30-31, 38, 43, 46, 51, 68, 81, 84, 99
American Safety Council, 19
American Society of Mechanical Engineers, 33, 38, 42, 71
American Society of Safety Engineers, 70-71
American Trucking Engineers, 71
Amm, James, 26
Amoco Production Co., 100, 103
Anderson, Paul, 108
Andrews, Andy, 93
A-1 Bit & Tool Co., 100
Ardmore, OK, 21
Argentina, 26, 83, 84, 100, 109
Arkansas, 13
Arlen, Richard, 53
Armstrong, John A., 97, 101, 107
Armstrong, L.D., 5
Association of Drilling Contractors, 46
Atlantic Richfield Co., 97, 100-101
Atlas Life Insurance Co., 5
Audit Bureau of Circulations, 96
Austria, 26

Baber, H.B., 82
Baden, W.A., 63
Badger, O.B., 62
Ballin, A.E., 79
Ball, John S., 107, 109
Barbour, Shirley, 75
Barit, A.E., 52
Barnett, Victor F., 48
Barnes, Wendell B., 83
Barnsdall Oil Co., 15
Barnsdall Research Corp., 61
Barnum, Don, 20
Baron, Thomas, 98
Bartlesville, OK, 42, 71, 80, 88, 96, 100

Bartlesville Research Center, 108
Bartlett, David, 97
Bartlett, Dewey, 97, 99-101
Barton Show Grounds, 3, 11, 12
Basic Science, Inc., 105
Batelle Co., 103
Bates, J.W., Jr., 97
Battisti, Jose M., 68
Bechtel Industries, 103
Beckstrom, R.C., 42
Beekly, A.L., 5
Beggs, OK, 48
Belcher, Page, 97
Belew, H.H., 97
Belgium, 40
Bell Helicopter, 99
Bellmon, Henry, 91, 97
Bennett, B.W., 109
Bentley, Frank, 66
Berliet Co., 96
Berry, Guy T., 63
Bethlehem Steel Corp., 24, 26, 70
Bethlehem Supply Co., 53, 66
Big Chief Drilling Co., 97
Bird, John W., 100
Blackman, Carl, 5
Blow, Almond, 53
Blythe Eastman Dillion & Co., 108
Boggess, H.W., 62
Bolivia, 65
Boos, Margaret, 100
Borden, O.V., 3, 5
Bourne, A.F., 5, 15
Bourque, A.V., 5
Bovaird, D.D., 82, 97, 101
Bovaird, George W., 71
Bovaird Supply Co., 15, 50
Bovaird, W. Merve, 15, 65
Boyd, W.R., Jr., 68
Boy Scouts, 33, 38, 52
Braden Construction Co., 6
Braden Industries, 103
Braden Steel Co., 63
Braden, W.R., 63
Bradley, H.P., 71
Bradshaw Oil and Gas Co., 15
Brann, C.C., 66
Brazell, Reid, 82, 109
Brazil, 69, 80, 100
Breene, Edmond C., 89
Brewster Co., 66
Bridenthal, D.A., 82
Bridgeman, Elizabeth A., 88
Bridgeport Machine Tool and Supply Co., 50
British West Indies, 26, 69
Brooks, J.H., 66
Brooks, Joan, 53
Brooks, Leslie, 3, 4, 11, 14, 48, 53, 60, 62, 63, 76, 82, 83, 96
Brown, A. Doug, 103

Brown, Bob D., 107
Brown, George O., 45
Brown, Kermit E., 103
Brownley, Claude B. III, 93
Bryant, Frank D., 63
Bryant, F.W., 5
Buchanan, D.E., 5
Buchner, C.E., 49
Buell, J. Garfield, 56
Bullard, E.F., 65, 82, 97, 101
Bunn, George P., Sr., 106
Bunn, George P., Jr., 101
Burdick, Virginia, 17
Burke, Edmund, 20
Burkey, Howard L., 66
Burks, Bill M., 109
Burlingame, Mark V., 109
Bush, Roy R., 70
By-Laws Committee, 63
Byles, Axtel J., 38, 51

Cabot, Goddrey L., 80
Cafe de Petrol, 53
California, 13, 49, 51, 73, 76, 100, 109
California Building, 33, 40, 65, 92
Calvert, F. Allen, 97
Calvert, F. Allen, Jr., 97, 101
Canada, 57, 80, 98-100, 109
Caravantes, Gloria, 105
Cassidy, William A., 89
Carter Oil Co., 66, 106
Caterpiller Tractor Co., 58, 95
Cavanaugh, Eddie, 18
C-E Natco, 105
Central Park, Tulsa, 11
Century Geophysical Corp., 70
Champion, John, 5, 53
Champlin Petroleum Co., 97
Charles R. Morse, Inc., 104
Chase, Frank L., 63
Chickasha Gas and Electric Co., 15
Chicken Farm wells, 10, 24
Chile, 69
China, 13
Chrysler Motor Co., 60
Cities Service Oil Co., 71, 101, 102, 108
Civitan Club, 18
Claremore, OK, 26, 80
Clark Brothers, 60
Clarke, George W., 105
Clark, Frank R., 100
Clark, L.G., 62
Cleers, Bernice W., 89
Cline, Irma, 89
Clinton, Fred S., 5, 6
Coffman, Charles L., 108
Collins, Don, 100
Colombia, 13, 26, 69
Colorado, 45, 100
Colorado School of Mines, 63, 101
Combustion Engineering Co., 102
Condon, Glen, 53
Connally, John, 93
Connecticut, 71
Connelly, W.L., 55, 71, 80, 83, 89
Continental EMSCO, 85, 89

Continental Oil Co., 103
Continental Pipeline Co., 97
Continental Supply Co., 72
Convention Hall, Tulsa, 3, 5, 6, 8, 9
Cook, Nell, 8
Cooley, H.M., 70
Collidge, Calvin, 22
Cooperative Club, 18
Cooper-Bessemer Corp., 73
Cooper, Fred E., 63
Cooper, Howell C., 63
Cooper, R.W., 62
Cordell, O.L., 5
Core Laboratories, 74
Cornelius, E.H., 5
Corrigan, Art, 103
Cosden and Co., 8, 9
Cosgrove, H.M., 63, 69, 70
Council of British Manufacturers of Petroleum
    Equipment, 77, 100
Cowan, Kirby, 48
Cox, Ralph F., 98
Cranston, L.E., 97
Crawford, Harry J., 80
Crawford, John M., 63
Creek Indians, 11
Critchlow, Walter, 21
Crooks, Stanley B., 100
Crosbie, John E., 55
Crystal Ballroom, Tulsa, 23, 48, 89, 100
Cuba, 69
Cubbage, T.J., 97
Cummings, Robert, 53
Cunningham, Bettye, 105
Cushing Oil Field, 3
Cushing, OK, 22
Czechoslovakia, 40

Daasch, F.J., 71
Dague, A.B.C., 5
Dalrymple, Dal, 48
Daniels, Lee, 97
Davidson, G.W., 100
Davis, Virginia, 92
Dawson, OK, 3
Deep Rock Oil Corp., 80
Denmark, 26
Department of Commerce and Mines, Mexico, 22
Department of Energy, 108
Desk & Derrick Club, 80, 88, 92, 96, 100, 105, 109
Devonian Oil Co., 15, 49
Dickeman, Raymond D., 104
Dickson Brothers, Inc., 104
Dillion, Letha, 80
Dimmick, Oppie, 70, 80
Doherty, Henry L., 32
Donald, James, 103
Donnell, J.C., II, 93, 97, 101
Douce, William C., 98, 107
Dover Corp., 100
Dowell, Inc., 70, 74, 86
Drake, Ed, 108
Drake, E.L., 9, 13, 19, 22, 23, 33, 36, 41, 48, 55, 63, 65, 72, 83, 88, 89
Dram, Ira H., 109
Dresser Industries, 66, 72, 92

Duperrey, Madame, 56
Duperrey, Maurice, 56
DuPont Co., 72, 105
Duvall Motor Co., 52
Duvall, William, 52
Dwyer, Martin C., 91, 92, 96, 100, 109

Earlougher, R.C., 108
Eastern Torpedo Co., 9
Eastman Oilwell Supply Corp., 62
Eastman Oilwell Surveying Co., 74
Ecology Hall of Science Building, 98, 99
Ecuador, 26, 69
Edmondson, J. Howard, 83, 84
Egloff, Gustav, 55, 56, 58, 69, 79
Egypt, 13
Ehrhardt, Paul H., 14
Eisenhower, Dwight D., 83
Elkins, Lloyd F., 108
Elliott, John E., 96
Emery, Grace, 17, 18
Emery, Louis E., 17
Empire Companies, 42
Empire Refineries, 19, 22
Employers' Liability Assurance Co., 62
Engineering Laboratories of Tulsa, 61
Engineers' Society of Tulsa, 108
England, 13, 26, 40, 80
Enright, Mike C., 101, 107, 110
Erskine, H.S., 97
Espy, W.E., 5
Estes, Harry E., 100
Ethington, Donald, 103
Ethyl Corp., 109
Eugene Dietzgen Co., 61
Everett, C.T., 5
Exchange National Bank, 5, 26, 30
Executive Committee, 74, 81, 82, 92, 97, 101
Exploration Drilling Day Committee, 97

Fallen, Charles F., 5
Farmer, A.L., 4
Farrless, Benjamin, 68, 69
Federal Bureau of Mines, 22
Federal Oil Conservation Board, 32
Fell, Harold B., 49, 69
Fields, Dixie, 18
Filley, E.R., 5 Finance Committee, 69
First Lady of Petroleum Committee, 100
First National Bank & Trust Co., 82
Fisher, William A., 103
Flanagan, J.P., 55, 71
Fleming, J.A., 62
Florida, 109
Foamite-Childs Co., 13
Ford Motor Co., 86, 99
Foresman, Bob, 93, 105
Foster, C. Vernon, 103
Foster, Dean E., 71
France, 13, 26, 28, 40, 68, 73, 77, 80, 100
Franchot, N.V.V., 45, 48
Franklin Supply Co., 105
Franks Manufacturing Co., 54, 60, 73
Fraternal Order of the Knights of the Derrick, 12
Fred E. Cooper Co., 63
Frick-Reid Supply Co., 15, 49, 50

Fries, Herbert C., 105, 109
Frisco Railroad, 68
Fritschen, Herman, 108
Fritzel, Max J., 103
Fuqua, B.B., 103

Gallagher, Ralph W., 71
Gallman, William, 100
Galloway, Bob L., 98
Gammelgard, P.N., 98
Gannaway, W.R., 97
Gardner, James H., 5, 57
Gardner Petroleum Co., 5, 57
Gardner, W.H., 55, 56
Garland Airport, 29
Garret-Airsearch, 99
Gaso Pump and Burner Co., 50
Gates Hardware Co., 50
Gauine, Katherine, 8
Geffen, T.M., 103, 108
General Electric, 19
General Motors, 17, 23, 59, 66, 69, 92, 94, 99
Geolograph Co., 105
Geophysical Research Corp., 61, 105
George, Charles J., 109
George E. Failing Co., 66
Georgia, 109
Germany, 13, 56, 77, 80, 95
Getty, J. Paul, 106
Getty Oil Co., 108
Gibbons, J. Barr, 4, 5, 20, 21, 24, 25-29
Giddens, Paul, 89
Gilles, Hal, 18
Gillespie, Bart W., 109
Gill, Joe, 48
Gilmer, F.P., 58
Ginter, R.L., 5
Glaser, David L., 97
Glass, J. Wood, 46
Glasscock, Frank, 5
Glennpool Oil Field, 3
Glenn Pool, OK, 49
Goddard, H.H., 5
Goebel, Art, 23
"Golden Driller," 92, 112
Gordon, A.W., 45
Gore, Thomas P., 37
Graham, N.R., 5
Graham, O.W., 93
Grainard, G.R., 101
Gray, Frank, 49
Green, William G., 63
Greenwood, Morgan, 103
Greer, Damon, 107
Gregerman, Ira, 108
Gregory, Dallas, 42
Gregory, June, 80
Grey, W.H., 9
Griffis, Keith, 99
Grimes, Otha H., 109
Gruber, H.R., 48, 53
Guatamala, 13, 26
Guerrero, E.T., 98, 103, 109
Guiberson, W.B., 17
Gulf Oil Corp., 62, 71, 81, 108
Guthrie, E.B., 5

Gwynne, R.D., 5

Hacke, Maxine, 92
Halliburton Building, 72
Halliburton Oilwell Cementing Co., 74, 85, 99, 108
Halliday, W.B., 98
Hall of Science Building, 50, 55, 62, 71, 73-74, 79-80, 88, 108, 110
Hall of Science Committee, 73, 79
Hamill, Ruth, 100
Hamilton, W.R., 5
Hamline University, 89
Hamon, Jake L., 82
Handwerk, Glenn E., 103
Hanford Airlines, 52
Hanlon, Edward I., 89
Hanlon, M.P., 55
Hanson, Clifford, 97
Hara, James E., 107
Hardin, E.P., 101
Hardy, C.M., 63
Hardy, Summers, 5
Harnett, Joseph B., 103
Harrigan, Barney E., 26
Hartman, Thomas J., 5
Haskell, Frank A., 71-72
Haskell Institute, 18, 20
Hayner, J.M., 5, 6
Hays, Roy M., 97
Hayward, John T., 61
H.C. Price Co., 97
Heacock, B.C., 58
Healdton Oil Field, 3
Hedrick, Paul, 48
Heggem, Alf G., 5, 15, 21, 25, 27, 53, 65, 68
Helmerich & Payne Drilling Co., 80, 97
Helmerich, W.H., 80, 82, 92
Helmerich, W.H. II, 97, 101, 107
Hembree, Mary, 100
Hendee, Robert W., 100
Henderson, Patricia, 100
Heston, J.E., 97
Hewgley, James M., 93
Hidell, Frances C., 105
Highland Body Manufacturing Co., 19
Hill, R.H., 86
Hinderliter, Frank, 5, 9-10, 15, 21, 25, 46, 53, 65, 89, 92
Hinderliter Tool Co., 50
Hissom, Wiley B., 89
Hochenauer, C.J., 80
Hodges, Hollis, 66
Holden; William A., 4-6, 15, 17, 21, 25, 46
Hollow, Rosaline, 8
Holmes, A.C., 5
Holmes, Fred, 48
Home Oil Co., 109
Hopkins, Howard, 96
Hormol, A.N., 96
Horrigan, Barney, 48
Hotel Tulsa, 11, 53, 66
H.O. Trerice Co., 105
Houchin, John M., 97, 101, 107, 110
Housing Committee, 83
Howard, George, 103
Hudson Motor Co., 52
Hudson, Rochelle, 53

Huff, Lyman C., 33
Huffman, R.S., 62
Hughes, George, Jr., 98
Hughes, Howard, Sr., 86
Hughes, I.H., 98
Hughes, Richard, 5
Hughes Tool Co., 74, 86, 109
Hull, B.E., 71
Humble Oil and Refining Co., 92, 97
Humes, R.P., 5
Hungerford, Clark, 68
Hurley, A.W., 5
Hurley Gasoline Sales Co., 66
Hurley, Patrick J., 32, 35
Hurst, R.E., 103

IBM, 93, 96, 103
Ickes, Harold, 45
Illinois, 13, 89
Independent Oil Co., 20
Independent Petroleum Association of America, 46, 69, 76
Independent Petroleum Association, Texas, 38
India, 13, 15
Indian Territory, 11
Indian Territory Illuminating Oil Co., 62
Institution of Petroleum Technologists, 46
Instruments, Inc., 93
Insull, Fred W., 15
International Delegates Committee, 49, 55
International Harvester Co., 50, 60, 72, 99
International Petroleum Exposition and Congress, 1923, 3-10; 1924, 10 15; 1925, 15-20; 1927, 20-24; 1928, 28-28; 1929, 28-31; 1930, 31-35; 1934, 36-46; 1936, 46-50; 1938, 50-56; 1940, 56-64; 1948, 65-74; 1954, 74-81; 1959, 81-89; 1966, 90-96; 1971, 96-101; 1976, 101-106; 1979, 106 109; evaluated, 110-112
International Petroleum Exposition, Inc., 5, 1, 15, 16, 21, 31, 33, 45
International Petroleum, Ltd., 57
International Petroleum Producers" Assn., 48
Interstate Compact Commission, 97
Iowa Petroleum Assn., 38
IPE Building, 91, 95-97, 99, 107
IPE Educational Building, 103
IPE Golf Tournament, 18, 21, 30, 35, 49
IPE Housing Committee, 97
IPE Theater, 109
Irizarry, Oscar B., 69, 80
Irwin, J.L., 82
Irvin, W.A., 47
Italy, 13, 26, 40, 80, 95

Jackson, Lewis B., 5, 6
James, William L., 66, 100
Japan, 13, 26, 55, 98
Jensen, Warren L., 101, 107
Jerome, Reed, 45
Jester, Beauford, 68
Joe Carson Post, American Legion, 40
Johnson and Fagg Engineering Co., 74
Johnston, Henry S., 22
John Zink Co., 108
Jones and Laughlin Co., 15, 66, 97
Jones, Jack D., 107
Jones, Vernon T., 109
Judd, Jack, 97, 101, 107

Junior Chamber of Commerce, Tulsa, 18, 33, 46, 48, 53, 63

Kansas, 17, 21, 45, 54, 58, 63, 71, 73, 86
Kansas Building, 92
Kaye, Emby, 57
Keating, A.F., 82, 83
Keener Oil Co., 97
Keller, George V., 103
Kelsey, Dana H., 71
Kemm, James O., 99, 107, 108
Kennedy, Ann, 8
Kennedy, Joseph B., 92
Kennedy, L.E., 5
Keplinger, C.H., 82
Keplinger, H.R., 80, 108
Kerr, C.W., 22, 26, 30
Kerr-McGee Oil Industries, Inc., 93, 97, 103, 104
Kerr, Robert S., 53
Kettering, C.F., 17, 69
Kewanee Oil Co., 97
Kidder, H.J., 103
Kidd, Robert L., 97
Kilmer, Jim, 108
Kimball, Kent W., 79
Kimball, K.K., 74
King, S.H., 3
Kinley Brothers, 43, 49, 50
Kinley, Floyd, 44, 49
Kinley, M.M., 44-45, 49-50, 86, 95
Kirchner, King P., 109
Kirkpatrick, E.L., 53
Kitchen, Margaret A., 89
Kiwanis, Tulsa, 13
Knights of the Derrick, 18
Knowles, Ruth S., 81
Koge, Reda Pump Co., 74
Kroll, Cornelius, 5
Krug, J.A., 67
Krusz, Harry, 48
Kryska, Terrace B., 103
KVOO, 25, 29, 38, 62
Kyle, E.H., 86

LaFortune, J.A., 53, 57, 66
LaFortune, Robert, 101
Lahn, G.C., 103
Lamont, Robert P., 31-32, 35
Landry, Art, 8
Lane Wells Co., 69
LaPlante, Roger, 100
Larkin, J.J., 34
Larkin Torpedo Co., 34
Lasater, R.M., 85
Latta, Robert, 92
Lauder, R.J., 62
Lauinger, P.C., 8, 53, 69, 106
Lauinger, P.C., Jr., 101, 107
Lawrence, K.M., 94
Lawson, Edward C., 49
Ledterman, R.L., 66
Lee, Larry, 53
Lemason, C.M., 5
Leonard, A.W., 15, 25, 29, 46, 48
Lerch, Frank J., 89
Leslie Brooks & Associates, 106

Leuty, Ben D., 109
Levorsen, A.I., 82
Lewis, W.L., 5
Lillie, Gordon W., 49
Lindbergh, Charles A., 23
Linden, Henry R., 104
Lindley, Joe R., 108
Lions Club, Tulsa, 18
Liquified Petroleum Gas Co., 5
Lockwood, G.O., 42, 71
Loftis, John L., 92
Long, I.G., 5
Looker, John C., 30
Louisiana, 900, 96, 100, 109
Louisiana Gulf Coast Exposition, 90
Lovejoy, John M., 46
Lundy, Roy, 68
Lyons, Charlton H., 76
Lyons, Paul, 97

MacArthur, Robert F., 15
Mack, P.H., 63
Maher, John, 108
Majestic Theater, Tulsa, 8
Manion, Jimmie, 49
Manning, Everett, 5
Mapes, Clarel B., 53
Marathon Oil Co., 93
Margay Oil Co., 81
Marland, E.W., 47, 53
Marland Oil Co., 19, 22, 47, 88
Marshall, John B., 74, 75
Marshall, Mrs. J.B., 76
Martin, George F., 49, 51
Martin C. Dwyer, Inc., 97
Marvis, N.B., 97
Matson, G.L., 5
Matthews, Bob, 108
Mattoon, Frank, 63
Maxwell, James L., 83-84
Mayer, T.F., 5
Mayo, Burch, 82
Mayo Hotel, Tulsa, 16, 18, 23, 25, 48, 66, 88, 100, 105
McBeth, Reid S., 46
McBirney, J.H., 5
McClure, H.O., 4-5
McComas, Murray, 108
McCormick, Julia, 105, 109
McDonald, George, 38
McDonnell-Douglas, 103
McDowell, R.W., 58, 66
McElroy, H.E., 5
McFarland, R.L., 5
McGee, Dean A., a92, 97, 101, 103-104
McGoldrick, J.K., 71
McGraw, James J., 5-6, 15, 18, 21, 25
McHardy, Bob, 108
McIntyre Airport, Tulsa, 23
McIntyre, Edward P., 6, 10-11, 15, 17-18, 20-21
McIntyre, James, 48
McKee, Calvin, 107
McKelvey, J.S., 5
McKinney, A.W., 82
McLerran, A.R., 103
McMahon, Charles L., 46, 55, 63
McMan Oil Co., 81

125

McReynolds, L.A., 98
McWilliams, J.R., 66
Melrose Oil Co., 49
Melton, W.A., 5
Meritt, H.C., 52
Mermudez, Antonio J., 68
Merriam, John F., 82
Messall, Flo, 105
Mexico, 13, 22, 26, 80, 100, 109
Meyers, Charles, 5
Michigan, 89, 109
Mid-Continent Oil and Gas Assn., 15, 19, 22-23, 33, 38, 53
Mid-Continent Oil and Gas Co., 26, 62, 81
Mid Continent Petroleum Corp., 58, 62, 66, 73
Mid Continent Petroleum Safety Council, 53-54
Mid Continent Supply Co., 84, 92
Miers, Shep, 70
Miller, E.B., Jr., 97, 101
Miller, Maurice, 108
Millinger, Larry, 108
Minger, J.E., 5
Minnehoma Oil Co., 106
Minnesota, 100
Miskell, P.M., 5
Missouri, 45, 49
Missouri Oil Men's Assn., 38
Mitchell, Gracie, 96
Mitchell, James A., 103
Mobil Exploration Service Center, 103
Moffitt, D.W., 5, 26
Moody, Nelson K., 49
Moore, E.H., 53
Moore, S. Donald, 103
Moran, John J., 106
Moran, Martin, 30
Morocco, 40
Mosier, Martain H., 48
Murray, C.M., 5
Murray, Johnston, 75-77, 80
Murtan, Frank B., 106
Muscovalley, J.N., 103
Museum of Petrology, 7
Myers, Barton T., 57, 74

Nassikas, John, 97
National Aeronautical Association
National Assn. of Independent Oil Producers, 9
National Petroleum Council, 81
National Stripper Well Assn., 46
National Supply Co., 13, 50, 72, 92
National Tank Co., 92-93
Natural Gas Assn., 83
Natural Gasoline Assn. of America, 46
Natural Gas Supply Men's Assn., 29
Needham, Riley B., 108
Netherlands, 80
Newblock, Herman F., 22
New Jersey, 29, 71, 89, 109
New Mexico, 96
New York, 23, 26, 69, 69, 71, 89
Nicholson, C.W., 83
Nippon Oil Co., 63-64
Nixon, Richard, 100
Noland, Michael C., 104
NOMADS, 63, 69, 80, 88, 100, 103, 105, 109

NOMADS International Headquarters Building, 67, 70, 71, 76
NOMADS IPE Committee, 100
North American Aviation, 95
Northeaster Oklahoma Stripper Well Assn., 46
Norwalk-Turbo, Inc., 104
Nunley, Perry L., 73

Oberfell, George G., 80
Odell, Carla, 108
Odom, Retha, 100
Offshore Day Committee, 97
Offshore Technology Conference, 107
Ohio, 72
Ohio Oil Co., 76
Oil Co. Buyers Group, 46
Oil Equipment and Engineering Exposition, 50
Oil Well Cementing Co., 72
Oilwell Supply Co., 50
Oil World Exposition, 50
Oklahoma, *passim*
Oklahoma A. & M. College, 38, 63, 76
Oklahoma Building, 25, 27, 31, 92
Oklahoma City Oil Field, 30
Oklahoma City, OK, 23, 42, 45, 51, 97, 100, 104, 109
Oklahoma Geological Survey, 20
Oklahoma Historical Society, 100
Oklahoma Inventors' Congress, 101
Oklahoma Military Academy, 24, 47, 53
Oklahoma Natural Gas Co., 62
Oklahoma Oil Co., 72
Oklahoma Petroleum Council, 99-101, 107-108
Oklahoma Power and Water Co., 62
Oklahoma State Unversity, 102, 110
Okmulgee, OK, 54
Old Timers Committee, 48, 53, 55, 63, 71, 82, 96, 100
Olson, Arthur O., 109
Orr, David, 108
Osage Indians, 3
O'Shaughnessy, I.A., 100
Otte, Carel, 104
Owen, J.P., 100
Owens-Corning Fiberglass, 103
Ozarks Chemical Co., 61

Paist, Stanley S., 103
Panama, 69
Pan American Petroleum Corp., 97, 99
Pape, C.H., 46, 53, 65, 82
Pariss, T.M., 5
Parker, G.C., 91
Parker, Ivy M., 96, 100
Parkersburg Rig and Reel Co., 86
Patton, Dan, 26, 30
Patton, Jim, 108
Patton, Mark S., 66
Pav, Mary E., 109
Pawley, Julian A., 103
Payne Petroleum Co., 109
Payne, William T., 109
Pease, F.T., 98
Peddycoart, L.R., 103
Peiffer, William H., 48, 49
Pennsylvania, 13, 18, 45, 48, 71-72, 80, 83, 89, 96, 100
Penny, T.A., 38, 47
Pennzoil Company, 71

People's Ice Co., 15
Permian Basin Oil Show, 90
Perry, E.R., 8
Perry, Neal V., 103
Peru, 13, 26, 69, 109
Peters, Ernesto, 83-84
Petroleos Mexicanos, 68[QL. Petroleum Bureau of Venezuela, 68
Petroleum Business Services, Inc., 89
Petroleum Council Historical Committee, 101
Petroleum Engineer Publishing Co., 50
Petroleum Equipment Suppliers Assn., 90
Petroleum Publishing Co., 97
Petroleum Safety Council, 43
Pew, J. Edgar, 17, 30
Pew, J. Howard, 96
Pew, Johathan G., 109
Pforzheimer, Harry, 104
Phillips, Frank, 23, 71
Phillips, Leon C., 59
Phillips Petroleum Co., 19, 71, 80, 88-89, 97, 103, 107
Pioneers of the Oil Industry Assn., 30, 34, 45
Plantation Pipeline Co., 96
Poe, Ronald L., 103
Poland, 13, 26, 34, 40
Ponca City, OK, 22, 103
Portable Gasoline Plants, Inc., 109
Porter, Frank M., 84
Porter, Hollis F., 5
Powers, H.R., 46, 65, 82
Powers, Jack, 53
Prairie Oil and Gas Co., 11, 15-16, 20-21
Pratt, Wallace E., 106
Price, Herold C., 97
Producers National Bank, 5
Production-Pipeline Day Committee, 97
Public Service Co., 9, 103
Purchasing Agents Assn., 46, 63
Pure Oil Co., 50
Pyzel, Daniel, 100

Raigorodsky, Paul M., 49
Raitz, C.H., 103
Ralph, Henry, 78
Ramseur, Fred H., Jr., 107, 108
Ramsey, Asa E., 5
Randall, Lillian, 8
Raphel, S.J., 82
Ray, R.H., 74
Reading & Bates Co., 97, 103
Reagan, Francie, 109
Reagan, Ronald, 111
Red Cross Building, 39
Red Fork Pool (oil field), 11
Red Wing Shoe Co., 99
Reeder, Charles B., 98
Reeser, E.B., 30, 32, 71, 80
Refiners and Marketers Building, 33, 40
Refining-Petrochemical Day Committee, 97
Reichl, Erich H., 104
Reilsing, Katherine, 8
Republic Aviation Corp., 86
Republic Supply Co., 66
Resource Sciences Corp., 103
Reynolds Metals Co., l72, 86
Rice, Frances Edgar, 96

Rice University, 101
Richards, E.A., 5
Richards, L.M., 109
Riley, Ralph C., 5
Ritchie, W.R., 5
Roach, Jack, 97
Roberts, George, Jr., 82, 99
Rockefeller, John D., Sr., 23, 26
Rockwell Manufacturing Co., 70
Roe, Harold E., 5
Roeser, Charles F., 46, 48
Rogers, C.B., 97
Rogers, Harry H., 26
Rogers, J.L., 97
Rogers, John., 53
Ronan, Isabel, 45
Roosevelt, Franklin D., 37
Rosser, L.G., 5
Ross, John, 11
Ross, Louis, 11
Rumania, 13, 26, 40, 44
Rumsey, I.A.P., 103
Russia, 13, 26, 100
Russian Oil Trust, 13

Safety Committee, 62
St. Francis Hospital Inc., 82
St. John's Hospital, Tulsa, 71
St. John Vianney Training School, 82
Salvation Army, 30
Sampson, R.E., 107, 108 Sanders, Andrew Jackson, 26
Sansing, Bill, 108
Sartori, J.A., 5
Saunders, J.B., 106
Savage, J.H., 62
Savit, Carl H., 103
Schaeffer, Inez Autey, 100
Scheopple, Andrew F., 58
Schermerhorn, William M., 63
Schleuter, W.A., 42
Schneider, R.T., 103
Schwab, Charles M., 24, 26
Scientific and Technical Building, 23, 25, 27-29, 42, 60, 92
Scientific and Technical Committee, 42, 82
Scotland, 80, 110
Scott, Cyrus McDonald, Jr., 100
Scott, Walter, 17
Secretary of Commerce (USA), 31, 35
Secretary of the Interior (USA), 45, 67, 97
Secretary of War (USA), 32, 35
Seismograph Service Corp., 48, 63, 86
Selby Oil and Gas Co., 53
Seminole Oil Field, 30
Seneca Oil Co., 109
Serebrovsky, A., 12, 13
Service Drilling Co., 97
Service, Willis J., 103
Shakely, J.L., 66, 82, 92, 97
Sharbaugh, Robert, 97, 101, 107
Sharp, Robert C., 69
Sharpe, J.E., 9
Sharrah, M.L., 109
Sheasley, Jacob, 23
Shell Oil Co., 62, 100
Sherry, William, 48, 82, 96, 100, 106
Shibley Engineering Co., 65

Sievert, O.M., 85
Silsbee, John, 49, 58
Simms, Dee W., 106
Simon, William, 108
Sinclair Consolidated Oil Co., 8, 22, 83, 89, 92
Sinclair, Harry F., 36
Sinclair-Prairie Oil Co., 55, 62, 71
Sisterville Field (West Virginia), 49
Skelly Oil Co., 15, 19, 22, 57, 73, 97, 107
Skelly Stadium, 32
Skelly, W.G., 15-18, 20-22, 24-26, 29, 30, 32, 33, 37, 38, 40, 46-48, 52, 53, 56-59, 62, 65, 66, 69, 74, 76, 77, 81, 84
Sloan, E.H., 34
Small Business Admin., 83
Smith, E.A., 97
Smith, George, 12
Smith, Glen J., 100
Smith, Harry, 5
Smith, Jack "Whispering," 53
Smith, L.E., 48
Smith, Turner C., Jr., 103
Sneed, Earl, 4, 10, 49
Sneed Oil Co., 80
Snell, Gene, 103
Societe Entrepose Gtm Pour Les Trebaux Petroliers Martimes, 100
Societe NMationale de Material Recherche L'Exploration, 77
Solar Aircraft Co., 85
Southern Hills Country Club, 53, 56, 80
Southwestern Assn. of Industrial Editors, 62
Southwest Research Institute, 103
Southwest Supply Co., 70
Spartan Dawn Patrol, 53
Spavinaw Lake, 13
Stalcup, H.M., 57
Standard Oil Co., 71
Stanolind Oil Co., 23, 66, 79
Staples, Oscar C., 9, 10, 15
Starr, Chauncey, 104
Staudt, John G., 70
Steiger, John, 101
Stephenson, J. Alexander, 23
Steward, C.R., 98
Stewart-Warner Corp., 105
Stewart, W.J., 48, 49
Stock, Paul, 96
Straight, Herbert, 71
Street, George, 77
Strozier, Marshall, 106
Suhn, Charles, L., 71
Sun Oil Co., 30, 89, 97
Swartz, W.H., 9
Swearingen, Eugene, 108
Swindell, Floyd, 48
Switzerland, 26

Tabor, George E., 63
Taher, Abdulahady H., 103
Tanner, N.E., 68
Taylor, Carroll C., 98
Texaco, 71, 93
Texas, 40, 45, 54, 68, 73, 83, 86, 89, 90, 93, 96, 100, 109
Texas Automotive Maintenance Assn., 38
Texas Building, 27, 31, 65, 72, 80, 92
Texas Company Refinery, 54

Texas Pipeline Co., 30, 54
Texoma Supply Co., 80
Thomas, Bill, 108
Thomas, Ross W., 106
Thompson, A.W., 82
Thompson, Ernest O., 83, 84
Thompson Oil & Gas Co., 15
Thornton, Charles, E., 103
Tidelands, 68
Tidewater Associated Oil Co., 57, 62, 71, 72
Tixier, M.P., 103
Tokheim, J.J., 60
Towl, Forrest, 63
Tracey, O.V., 109
Trans-American Pipeline Co., 71
Transcontinental Oil Co., 20
Trapp, M.E., 12
Tret-O-Lite Co., 66
Trees, Romona Marcella, 12, 17
Tri-State Building, 80
Tucker, E.T., 5
Tulsa Assembly Center, 100
Tulsa Building, 48, 63
Tulsa Camera-Record Co., 61
Tulsa Central High School, 45, 53, 69
Tulsa Chamber of Commerce, 4, 5, 13, 17, 18, 23, 38, 46, 53, 54, 56, 68, 74, 75, 81, 90
Tulsa Club, 8, 49, 63, 80, 90
Tulsa Country Club, 18, 21, 31, 35
Tulsa County Fair Board, 65
Tulsa Exposition & Fair Corp., 74
Tulsa Free Fair Board, 21, 24
Tulsa Frisco Railway Station, 3, 6
Tulsa International Airport, 109
Tulsa Municipal Airport, 29, 53, 56, 80
Tulsa, OK, *passim* Tulsa Petroleum Club, 80
Tulsa Philharmonic, 80
Tulsa Public Schools, 62
Tulsa Purchasing Agents Assn., 3
Tulsa Radio & Electric Co., 19
Tulsa Rotary Club, 56
Tulsa State Fair, 3, 91
Tulsa State Fairgrounds, 91, 112
Tulsa Testing Laboratories, 74
Turner, Mary Lee, 105
Turner, Roy J., 68, 69

Udden, J.A., 5
UNESCO, 73
U.S. Bureau of Mines, 54
U.S. Chamber of Commerce, 12
U.S. Department of Commerce, 107
U.S. Geological Survey, 12, 20, 101
U.S. Junior Chamber of Commerce, 48
U.S. Steel Corp., 47, 60, 69
Universal Oil Products Co., 33, 35, 58, 69
University of Oklahoma, 62, 63, 71, 110
University of Tulsa, 8, 18, 322, 42, 53, 63, 82, 102, 103, 105, 109, 110
University of Wyoming, 101
Uruguay, 69

Valerius, M.M. 5
Vandever, W.A., 5, 6
VanTine, John W., 55
Veale, C.H., 58

Veasey, James, A., 63, 71
Venezuela, 13, 26, 34, 69, 80, 100
Vensel, Dorothy, 8, 12
Venturtek International, Ltd., 104
Vickers, E.C., 53

Walker, J.M., 69
Wall, C.I., 106
Wallak, Hazel, 8
Waller, Michael, R., 109
Ward, Robert W., 103
Warren, C.M. 76
Warren, J.S., 5
Warren Petroleum Co., 53, 58, 66, 81, 92
Warren, William K., 81-84, 90-93, 96, 106, 111
Washecheck, Paul H., 103
Watkins, George, 32
Waukesha and Orenda, Ltd., 99
Way, W.B., 29, 31, 33, 35-40, 46-48, 53, 57, 58, 65, 66, 69, 71, 74, 76, 80, 82, 91
W.C. Norris Co., 50
Weatherby, B.B., 73, 79
Weatherly, Albert, 45, 53
Webber, Charles E., 82
Weber, George H., 96
Welch, Van S., 96
Welch, W.M., 5
Wertzberger Derrick Co., 15
Wertzberger, D.D., 15
Westby, G.H., 79
Wheatly, Charles, 106
White, Carl F., 73
Whiteside, Allan, 5

Wiet, E.H., 5
Williams, C.V., 62
Williams, David R., 98
Williams, Luther, 62
Williams, S. Miller, 89
Will Rogers Memorial, 80
Willson, C.O., 63
Wilson, Grace, 18
Wilson, Riley, 93, 94, 107
Wilson, Wallace D., 80, 81
Windsor, Frederick Ernest, 17
Wirth, Alfred, 13
Wittig, D.E., 103
Wolf, Frank, 55
Woman of Achievement Committee, 80
Wood, Phil, 108
Woolaroc Museum, 80
World Petroleum Congress, 68, 110
Wright, Joe R., 98
Wright, Lilly E., 105
Wright, M.A., 97, 101
Wyoming, 97

Yost, F. Randolph, 97-101, 106, 107
Young, J.C., 62
Youngsters of the Oil Industry Committee, 63
Yuba Heat Transfer Division, 83
Yunker, William E., 53

Zarrow, Henry H., 107
Zeppa, Joseph, 100
Zevada, Manuel J., 22
Zink, John, 5

A000013026608